SpringerBriefs in Architectural Design and Technology

Series editor

Thomas Schröpfer, Singapore University of Technology and Design, Singapore, Singapore

Understanding the complex relationship between design and technology is increasingly critical to the field of Architecture. The SpringerBriefs in Architectural Design and Technology series aims to provide accessible and comprehensive guides for all aspects of current architectural design relating to advances in technology including material science, material technology, structure and form, environmental strategies, building performance and energy, computer simulation and modeling, digital fabrication, and advanced building processes. The series will feature leading international experts from academia and practice who will provide in-depth knowledge on all aspects of integrating architectural design with technical and environmental building solutions towards the challenges of a better world. Provocative and inspirational, each volume in the Series aims to stimulate theoretical and creative advances and question the outcome of technical innovations as well as the far-reaching social, cultural, and environmental challenges that present themselves to architectural design today. Each brief asks why things are as they are, traces the latest trends and provides penetrating, insightful and in-depth views of current topics of architectural design. Springer Briefs in Architectural Design and Technology provides must-have, cutting-edge content that becomes an essential reference for academics, practitioners, and students of Architecture worldwide.

More information about this series at http://www.springer.com/series/13482

Thomas Schröpfer · Sacha Menz

Dense and Green Building Typologies

Research, Policy and Practice Perspectives

Springer

Thomas Schröpfer
Singapore University of Technology
and Design
Singapore, Singapore

Sacha Menz
ETH Zürich
Zürich, Switzerland

ISSN 2199-580X ISSN 2199-5818 (electronic)
SpringerBriefs in Architectural Design and Technology
ISBN 978-981-13-0712-6 ISBN 978-981-13-0713-3 (eBook)
https://doi.org/10.1007/978-981-13-0713-3

Library of Congress Control Number: 2018945064

Printed on acid-free paper

This Springer imprint is published by the registered company Springer Nature Singapore Pte Ltd.
The registered company address is: 152 Beach Road, #21-01/04 Gateway East, Singapore 189721, Singapore

Foreword: Propagating Singapore's Alternative Modernity

It is 10 AM on Saturday 24 March 2018 in the 'pod' auditorium on the 16th floor of Singapore's National Library. A large and expectant audience of assorted architects, artists, poets, sociologists and intellectuals is gathering. They have come to hear William Lim give a lecture celebrating the bestowal of this year's Singapore Institute of Architects Gold Medal upon him.

The lecture title appears on the screen in large bold letters: 'The Future is Now!' it announces. The audience settles down as the Chair and Institute President, Seah Chee Huang, recounts the many inter-twined threads of Lim's career as architect, writer, activist and cultural impresario.

On stage, Lim adjusts his mic, consults his notes and begins in characteristic academic style. As his lecture gathers steam, he builds a compelling case for the role of architecture in grappling with the multifarious challenges of contemporary urbanisation in Asia. He patiently tabulates the negative symptoms—land-grabbing, real estate speculation, displacement of the urban poor, destruction of culture and environmental degradation—and challenges his audience to address them in their respective day-to-day professional lives. His own position clear: 'starchitecture' cannot help, and local built examples by Zaha Hadid and Daniel Libeskind are summarily dismissed.

Those in the audience expecting reflection on the back-catalogue of Lim's own built work are growing restless. As the clock ticks past the half-hour mark, it becomes clear that there will be no slides on the Golden Mile or People's Park complexes, those audacious and famed mixed-use megastructures of the 1970s that he designed with colleagues Tay Kheng Soon and Koh Seow Chuan. It is no surprise that the post-lecture discussion turns to those projects. One after another, audience members reflect on the built work of Lim and colleagues of his generation such as Tay and Koh, as well as Tan Cheng Siong and Ken Yeang. Projects such as Siong's Pearl Bank Apartments, Tay's Dairy Farm Estate, and his development guide plan for Kampung Bugis, and Yeang's more recent work (including the National Library building in which the event is taking place), feature strongly.

Discussants, sociologist Kwok Kian Woon and architect Richard Ho, join Lim on stage to sum up. They skilfully capture the mood of the auditorium, and try to square Lim's future-oriented rallying speech with the reflective atmosphere in the room. Lim, Tay, Koh, Tan and many others of that generation, Kwok suggests, gave shape to a unique legacy of tropical architecture and city-making. This legacy was based on a productive tension that embraced both modernisation's progressive drive, technological optimism and universalism, as well as a postcolonial spirit for self-determination, cultivating cultural identity, and appreciation of tropical environmental conditions. This heady mix—implicating science, technology, architecture, urbanism, poetry, art and cinema—shaped an alternative modernity that inspired a wholly new city vision. The relatively short history of postcolonial city-making at that time—Chandigarh, Brasilia, Canberra and Japan's metabolism —was instructive. But early projects like the Golden Mile and the Kampung Bugis development guide demonstrated, for the first time, the possibilities of a densely populated, global, mixed-use city that was at ease in the tropics.

The future may be now, Kwok and Ho seemed to be saying, but it draws upon a rich legacy of innovation, creative risk-taking and appreciation for the nuances of culture and situation that Singapore's early alternative modernists shaped.

What, then of the now and the future? The alternative modernity legacy continues to inspire new projects, such as initiatives to document built work of the postcolonial period, efforts to capture the oral history of architects, engineers and actors of that era, and student architectural design experiments at Singapore University of Technology and Design (SUTD) and National University of Singapore (NUS) on 3-dimensional, high-density and bioclimatic forms. It might also be argued that many of the mainstream policies and visions of government planning agencies—in their embrace of non-motorised transport options, landscaping for urban spaces and high-rises (LUSH) and water-sensitive urban design, for example—build upon the possibilities of an alternative modernity.

This book can be counted among such projects too, expanding as it does the ecological and programmatic possibilities of high-density urbanism. The book emerges from research conducted by the 'Dense and Green Building Typologies' team at the Future Cities Laboratory (FCL). The team is composed of architects, academics, scientists and ecologists from ETH Zürich and SUTD, and in this book offers a precise demonstration of the FCL mission to conduct 'transformative research'.

The term 'transformative research' is most clearly articulated by the German Advisory Council on Global Change (WBGU) in its flagship report 'World in Transition: A Social Contract for Sustainability' (2011). In that report, the author team outlines what they call 'the great transformation' necessary to shift from the fossil fuel-based economic system towards a sustainable, low-carbon future. The great transformation is an ambitious enterprise, matched only, in the eyes of the WBGU, by two other events in world history for scale and impact: the neolithic revolution which saw the rise of agricultural society; and the industrial revolution that underpinned the rise of planetary urbanisation over the past two centuries (2011, 81). Other scholars agree. The Stockholm Resilience Centre team, for example, has diagnosed what it calls 'the great acceleration' of almost all demographic and economic indicators matched by a sharp deterioration of indicators of environmental health since the beginning of the industrial revolution.

For the WBGU, the great transformation involves ushering in fundamentally new forms of production, consumption and lifestyle in order to 'reduce greenhouse gas emissions to a minimum in the coming decades (decarbonisation of the energy systems and establishment of low-carbon societies), to minimise the scarcity of essential resources (above all land, water, strategic mineral resources) through major resource efficiency increases, and to avoid abrupt changes within the Earth system (tipping points), through economic and development strategies which take the guard rails of the Earth system (planetary boundaries) into account' (WBGU 2011, 81).

Transformative research is one of the 'instruments of interdisciplinary research' (WBGU 2011, 351) that is necessary to address such challenges. It spans discipline- and system-based research and encourages a cross-fertilisation between the two. It emphasises engagement 'with society, the economy, and politics', while considering 'global usability', 'intercultural transferability' and potential 'rebound effects and 'path dependencies'. In short, transformative research aspires to actively steward, guide and curate the great transformation processes in responsible ways.

By gathering together this rich body of research material, the 'Dense and Green' team adopts a nuanced stance: on the one hand, sensitively building upon the possibilities that early Singaporean architects, artists and intellectuals articulated; and at the same time, engaging with regional and global debates concerning sustainability and urbanisation. In this sense, the book could well be an instrument of transformative research, that gives substance to Lim's call.

Singapore, Singapore Stephen Cairns

Preface

The following book chapters are based on the various contributors' presentations at the 'Dense and Green Building Typologies: Architecture as Urban Ecosystem' symposium that took place at the Singapore Urban Redevelopment Authority (URA) on 30 August 2017. The event was organised in the context of the Future Cities Laboratory Dense and Green Building Typologies research project. It brought together important stakeholders, including researchers, policy makers, planners, developers, architects and landscape architects, to discuss how dense and green buildings in Singapore and beyond can contribute to developing compact yet highly liveable future cities.

Dense and Green Building Typologies Project

Dense and Green Building Typologies is a 5-year research project of the Future Cities Laboratory (FCL), established by the Swiss Federal Institute of Technology Zurich (ETHZ) and the National Research Foundation Singapore in collaboration with key academic partners including the Singapore University of Technology and Design (SUTD). FCL studies sustainable future cities through science, by design and in place, with its High-Density Mixed-Use Cities Scenario developing new integrated planning paradigms, research methodologies and implementation processes to support higher population densities, higher standards of environmental sustainability and enhanced livability. As part of that scenario, Dense and Green Building Typologies investigates systematically the environmental, social, urban, architectural and economic benefits of large buildings with integrated green spaces in high-density contexts through a series of international in-depth case studies, including in Asia, Europe and the Americas.

Singapore, Singapore — Thomas Schröpfer
Zürich, Switzerland — Sacha Menz

Acknowledgements: Dense and Green Building Typologies Research Team

Prof. Dr. Thomas Schröpfer is Professor of Architecture and Sustainable Design (ASD) at the Singapore University of Technology and Design (SUTD) and Principal Investigator of the Dense and Green Building Typologies project at the Singapore-ETH Centre Future Cities Laboratory (SEC-FCL). He began his academic career at Harvard University, where he was appointed Assistant Professor of Architecture in 2004 and Associate Professor of Architecture in 2008. He was named Full Professor after joining SUTD in 2011, where he became Associate Head of Pillar of ASD, Co-Director of the SUTD-JTC I^3 Centre and Director of the University's award-winning Advanced Architecture Laboratory. He held visiting professorships at the Massachusetts Institute of Technology, the Swiss Federal Institute of Technology and the National University of Singapore. In 2015, he was appointed member of the SEC-FCL Steering Committee. His book publications have been translated into several languages and include *Dense + Green: Innovative Building Types for Sustainable Urban Architecture* (2016), *Ecological Urban Architecture: Qualitative Approaches to Sustainability* (2012) and *Material Design: Informing Architecture by Materiality* (2011). He is the recipient of prestigious awards and recognitions, including the Singapore President's Design Award, the German Design Award the and the Asia Education Leadership Award. His research and design projects have been exhibited at important international venues, including the Venice Architecture Biennale and the World Congress of Architecture.

Prof. Sacha Menz is Professor of Architecture and Building Process at the Swiss Federal Institute of Technology (ETH) Zürich. He has served as the Dean of the Department of Architecture, and has been leading the Institute of Technology in Architecture (ITA) for over 10 years, where he designed and coordinated the building process of the Arch_Tec_Lab. He also established the Master of Advanced Studies (MAS) Programme in Architecture, Real Estate and Construction, as well as several CAS (Certificate of Advanced Studies) courses at ETH. He held the position of Deputy Director at NCCR (National Centre of Competence in Research) in Digital Fabrication and was a guest professor at South-Eastern University in Nanjing, China. Trained in Architecture at ETH with a master's degree, Sacha co-established sam Architects and Partners (SAM) in Zürich/Switzerland. SAM designed the award-winning»Vorderer Sternen« Building in Zürich. He is member of several boards such as the Architectural Board of Consultants for the City of Ostfildern, Germany; Board of Directors of Swiss Engineers and Architects Association (SIA), Reviewing Board for Clusters of Excellence of the DFG (German Research Foundation). He has authored academic publications including *Three Books on the Subject of Building Process* and *Public Space Evolution in High-Density Living in Singapore* and actively contributes in international conferences and architectural juries.

Dr. Michelle Yingying Jiang is a postdoctoral researcher and the Project Coordinator of the Dense and Green Building Typologies project. She investigates the post-occupancy performance of dense and green buildings in terms of social benefits, including user perception and preference, as well as dense and green buildings' influence on health and comfort. Her research interests include architectural history, space design and the interrelations between the physical quality of space and human behaviour. Prior to her current position, Michelle was part of Module X: Housing at the Future Cities Laboratory as a postdoctoral researcher, analysing socio-spatial qualities of elevated public spaces in public housing in Singapore. She also participated in the Smart Living Lab at the École Polytechnique Fédérale de Lausanne (EPFL), where she drafted the programme on building space flexibility and usability. Michelle graduated from Zhejiang University of Technology in 2004 with an Engineering Bachelor Degree in Architecture Design. She obtained an Master's Degree Engineering in Architecture, History and Theory from Shenzhen University in China in 2008. In 2012, she completed her PhD thesis on building flexibility and adaptability at the Architecture Department of The University of Hong Kong.

Richard Belcher's role at the Future Cities Laboratory is quantifying the environmental and economic value of greenery in dense and green building typologies using post-occupancy data. He holds a Master of Science in Environmental Management from the National University of Singapore (NUS), where he quantified how neighbourhood green space influences the selling price of public housing. He also has a BSc (hons.) in Forestry from Bangor University, Wales. Prior to his role at FCL, Richard worked as an environmental consultant in Singapore, conducting pollution control studies, Environmental Impact Assessments (EIAs) and emissions inventory studies.

Emek Erdolu's research at the Future Cities Laboratory examines the variations in quantum and design configurations of green components on and around the high-density buildings as patterns of provision and connectivity within larger urban green systems. His broader research interest is the relationship between digital technology and design thinking in the context of emerging computational tools and its implications for architecture and urban design. Prior to FCL, Emek has worked for 7 years in the US, China and Singapore on various architecture, landscape and urban design projects and he has taught as a studio instructor at Bilkent

University in Turkey. He holds a Master of Urban Design with Distinction from University of Michigan, Ann Arbor and a Diploma in Landscape Architecture and Urban Design from Bilkent University.

Mayank Kaushal is an architect, urban designer, researcher and an avid writer with sustainable design expertise. He has worked on diverse types in the public and private sector, spanning various cultural and climatic regions across Asia. Over the years, he was involved in a number of award-winning projects that celebrate sustainability by qualitatively and quantitatively sculpting environmental resilience. He is constantly working on combining research with practical applications. He has contributed to the Masters programme as a juror for studio presentations at the National University of Singapore and through his recent editorial work on 'SG3—Decoding Sustainable Urbanism, Singapore'. He is an advocate rethink his current role as a researcher at Future Cities Laboratory focuses on understanding qualitative and quantifiable design benefits of dense and green building typologies.

Prashanth Raju is a researcher at the Future Cities Laboratory of the Singapore-ETH Centre. As part of the Dense and Green Building Typologies project, his research examines the variations in design configurations of green components on high-density buildings and analyses their design benefits in architecture and urban scale. His broader research is on the integration of landscape infrastructure within the built environment using design strategies that strengthen the negotiation and interaction between them. Prior to FCL, he has worked as an architect for two years in India where he was involved in residential, commercial and master planning projects. Prashanth holds a dual Master of Urban Planning and Master of Urban Design degree from the University of Michigan, Ann Arbor, where he was awarded for his work on accessibility and affordability of design. He also holds a Bachelor of Architecture from MEASI Academy of Architecture, India.

Thibault Pilsudski's role at the Future Cities Laboratory (FCL) is investigating maintenance aspects of greenery in high-rise buildings as well as analysing them in terms of Green Plot Ratio (GNPR) as an indicator for density of greenery. Within the larger urban context, he also examines the impact of dense and green buildings in relation to urban density and accessibility to landscape spaces. Prior to FCL, he has worked at the Ministry of National Development Centre for Liveable Cities in Singapore on various urban projects such as the pedestrianisation of Orchard Road. Thibault holds a Master of Urban Studies and Public Policy degree from Sciences Po Paris and a Master of Urban Planning degree from the National University of Singapore.

Jonathan Koon Ngee Tan is a researcher currently working on quantifying the thermal benefits of urban greenery using infrared thermography, image classification and statistical modelling. During his time with the Botany Laboratory at the National University of Singapore, he surveyed secondary forests to detail the population dynamics of exotic tree species and their native ecological analogues and conducted nursery experiments to investigate the responses of exotic tree species along a nutrient gradient. His larger research interest is the functional benefits and conservation of nature. He holds a Bachelor of Science in Environmental Studies (Biology Specialisation) degree from the National University of Singapore.

Ester Suen's work aims to understand the benefits of greenery on buildings in Singapore with regards to biodiversity and microclimate. Her broad interests include urban ecology, sustainability and conservation. Specifically, she is interested in understanding the socioecological relationships between plants, animals and people in urban areas. She has previously worked in a forest ecology and restoration laboratory at Yale-NUS, where she studied the functional traits of plants in relation to their environment. She holds a Bachelor of Science in Wildlife Biology and Conservation from the University of Guelph in Canada.

Contents

Contributors

Manuel Der Hagopian G8A, Singapore, Singapore

Peter Edwards ETH Zürich, Zürich, Switzerland

Christophe Girot ETH Zürich, Zürich, Switzerland

Richard Hassell WOHA, Singapore, Singapore

Cheong Koon Hean Housing and Development Board, Singapore, Singapore

Ng Lang Urban Redevelopment Authority, Singapore, Singapore

Nicholas Mak SLP International Property Consultants, Singapore, Singapore

Siew Man Kok MKPL Architects, Singapore, Singapore

Yeo Meng Tong National Parks Board, Singapore, Singapore

Sacha Menz ETH Zürich, Zürich, Switzerland

Franklin Po Tierra Design and PODesign, Singapore, Singapore

Thomas Schröpfer Singpore University of Technology and Design, Singapore, Singapore

Ryan Shubin Ramboll Studio Dreiseitl, Singapore, Singapore

Henry Steed ICN Design, Singapore, Singapore

Khoo Teng Chye Centre for Liveable Cities, Ministry of National Development, Singapore, Singapore

Megan Walters JLL Asia Pacific, Singapore, Singapore

Ken Yeang T. R. Hamzah & Yeang, Kuala Lumpur, Malaysia

Tan Shao Yen CPG Consultants and Board of Architects, Singapore, Singapore

Acronyms

BCA	Building Construction Authority
CLC	Centre for Liveable Cities
FCL	Future Cities Laboratory
HDB	Housing & Development Board
JTC	Jurong Town Corporation
LEAF	Landscape Excellence Assessment Framework
LUSH	Landscape for Urban Spaces and High-rises
NParks	National Parks Board
NUS	National University of Singapore
PUB	Public Utility Board
SUTD	Singapore University of Technology and Design
URA	Urban Redevelopment Authority

Chapter 1
Dense and Green Building Typologies: Architecture as Urban Ecosystem

Thomas Schröpfer and Sacha Menz

1.1 A New Paradigm for High-Density Liveable Cities

Contemporary architecture and urban design practice in Singapore is increasingly exploring the integration of green spaces in buildings, producing new typologies for high-density environments that include public spaces, extensive sky terraces, sky bridges, vertical parks, roof gardens and other 'green' components. Combinations of all these, often applied to mixes of residential, civic and commercial programmes, conjoin at times to produce 'vertical cities' in which the building section becomes part of larger urban ecosystems such as parks, gardens and river networks. Density and sustainability here are not seen as contradictory but rather as mutually dependent and synergistic (Fig. 1.1).

Dense and Green Building Typologies—a 5-year research project of the Future Cities Laboratory, established by the Swiss Federal Institute of Technology Zurich and the National Research Foundation Singapore in collaboration with key academic partners including the Singapore University of Technology and Design—explores how these developments can support higher population densities, higher standards of environmental sustainability and enhanced livability in Singapore and beyond.

1.2 Why Singapore?

A number of fortuitous and strategic conditions have made Singapore a leader in dense and green typologies. Projects like WOHA's School of the Arts (completed 2010) and PARKROYAL on Pickering (completed 2013) have led the way on the design side with visionary leadership. Asia, in general, and Singapore, in particular, have great potential for the further exploration of dense and green as well as liveability principles. The breathtaking scale of many of the new developments here captures the attention of politicians and developers and the imagination of its citizens.

T. Schröpfer and S. Menz, *Dense and Green Building Typologies*, SpringerBriefs in Architectural Design and Technology,
https://doi.org/10.1007/978-981-13-0713-3_1

Fig. 1.1 WOHA, PARKROYAL on Pickering, Singapore, 2013, sky gardens. *Credit to* Patrick Bingham-Hall

As a small island state with limited land and natural resources, 'green' agendas have guided Singapore's developmental approach, even before the term became the buzzword commonly used today. Singapore's former Prime Minister, Mr. Lee Kuan Yew, had the vision of making Singapore a 'Garden City'. It was a revolutionary concept at that time because nobody talked about 'going green' or climate change (Ng 2012).

Since then, Singapore has recognised the importance and benefits of a green environment. Even as the city state embarked rapidly on industrialisation and urbanisation programmes to provide jobs and housing for its people, the environment was high on the Government's agenda. Today, Singapore's vision has evolved from 'Garden City' to 'City in a Garden'. This concept is seen to strengthen Singapore's brand as a distinctive, liveable city. The new vision also embodies the ideas of conserving and nurturing biodiversity in the urban context, an area where Singapore has contributed scientifically through the Singapore Index (for urban biodiversity).

As Singapore's population continues to grow and with limited land available, developing a compact city with extensive greenery and highly liveable environments, will continue to be an important strategy. Singapore has identified three key strategies for further pursuing its 'City in a Garden' vision: using pervasive greenery from the ground to the facade and rooftops of buildings, infusing biodiversity into urban landscapes and fostering of community involvement as active participation, ownership and pride among the community are seen as factors that will sustain the 'City in a Garden' vision (Poon 2012).

Since the early 2000s, Singapore has pursued a number of research and small demonstration projects that explored that integration of green spaces in buildings. These projects led to a number of policies and initiatives such as GFA Exemption for Communal Sky Terraces, GFA Exemption for Communal Planter Boxes, Skyrise Greenery Scheme, Landscape for Urban Spaces and High-rises (LUSH), as well as the Landscape Excellence Assessment Framework (LEAF). These have been instrumental for the subsequent experimentation with dense and green building typologies.

Many developments in Singapore now state selling points such as 'near a park', 'rooftop greenery' or 'vertical greenery'. The quantifiable benefit of such features is land value appreciation. At the same time, dense and green building typologies can also be seen as having a potential alleviating effect on land use competition as they are able to layer horizontal city functions vertically, thereby optimising land use in Singapore. At the same time, Singapore increasingly recognises unquantifiable benefits such as of living near rich biodiversity as well as the fact that being close to nature can result in improved physical and mental health and mitigate some of the negative effects associated with high-density urban environments (Tan 2012).

Further, pockets of green spaces are not just visual delights, nor simply for human use but can function as part of a larger urban ecosystem as well. For example, in Singapore's efforts to promote a greater biodiversity in its urban environment, dense and green building typologies can function as high-quality habitats for flora and fauna. Their combination with green spaces such as green corridors, parks, nature areas, and nature reserves can form an interconnected matrix that becomes part of a larger ecosystem. Dense and green building typologies can address environmental, social, architectural and economic aspects. As such, they can improve urban environments by mitigating negative effects of high density.

While the focus of *Dense and Green Building Typologies* is on Singapore, the research contributes knowledge about how such buildings can improve urban environments in other cities as well, steering the new dense and green paradigm from its current status to an integrated and tacit element of architecture as well as urban design and planning in the coming decades.

References

Ng, Lang. 2012. Singapore—Captial City of Vertical Green (interview). In *Architecture and Urbanism* Special Edition May 2012.

Poon, Hong Yuen. 2012. Singapore—Captial City of Vertical Green' (interview). In *Architecture and Urbanism* Special Edition May 2012.

Tan, Puay Yok. 2012. What does Vertical Green Mean. In *Architecture and Urbanism* Special Edition May 2012.

Prof. Dr. Thomas Schröpfer Thomas Schröpfer is Professor of Architecture and Sustainable Design (ASD) at Singapore University of Technology and Design (SUTD) and Principal Investigator of the Dense and Green Building Typologies project in the Future Cities Laboratory at the Singapore-ETH Centre (SEC-FCL). He began his academic career at Harvard University where he was appointed Assistant Professor of Architecture in 2004 and Associate Professor of Architecture in 2008. He was named Full Professor after joining SUTD in 2011, where he became Associate Head of Pillar of ASD, Co-Director of the SUTD-JTC I3 Centre and Director of the University's award-winning Advanced Architecture Laboratory. He held visiting professorships at the Massachusetts Institute of Technology, the Swiss Federal Institute of Technology and the National University of Singapore. In 2015, he was appointed member of the SEC-FCL Steering Committee. His book publications have been translated into several languages and include Dense + Green: Innovative Building Types for Sustainable Urban Architecture (2016), Ecological Urban Architecture: Qualitative Approaches to Sustainability (2012) and Material Design: Informing Architecture by Materiality (2011). He is the recipient of prestigious awards and recognitions, including the Singapore President's Design Award, the German Design Award and the Asia Education Leadership Award. His research and design projects have been exhibited at important international venues, including the Venice Architecture Biennale and the World Congress of Architecture.

Prof. Sacha Menz Sacha Menz is Professor of Architecture and Building Process at the Federal Institute of Technology (ETH) Zürich. He has served as the Dean of the Department of Architecture and has been leading the Institute of Technology in Architecture (ITA) for over 10 years, where he designed and coordinated the building process of the Arch_Tec_Lab. He also established the Master of Advanced Studies (MAS) Programme in Architecture, Real Estate and Construction, as well as several CAS (Certificate of Advanced Studies) courses at ETH. He held the position of Deputy Director at NCCR (National Centre of Competence in Research) in Digital Fabrication and was a Guest Professor at South-Eastern University in Nanjing, China. Trained in Architecture at ETH with a master's degree, Sacha co-established sam Architects and Partners (SAM) in Zürich/Switzerland. SAM designed the award-winning »Vorderer Sternen« Building in Zürich. He is member of several boards such as the Architectural Board of Consultants for the City of Ostfildern, Germany; Board of Directors of Swiss Engineers and Architects Association (SIA), Reviewing Board for Clusters of Excellence of the DFG (German Research Foundation). Sacha's research focuses on the future of building processes, the influences of economics in architecture and construction and on public space in the urban context. He has authored academic publications, Three Books on the Subject of Building Process and Public Space Evolution in High-Density Living in Singapore and actively contributes in international conferences and architectural juries.

Chapter 2
A City in a Garden

Ng Lang

Urbanism in Singapore is shaped by the desire to overcome the physical constraints of a small island city state. On a small island, the needs of both the city and the state have to be catered to. Above that, Singapore sustains one of the most competitive economies in the world and is a city ranked amongst the top in Asia for the quality of life. To meet the demands of these rather unique circumstances, Singapore has had to find its own ways to solve its problems, whether it is water management, public housing or transport. Over the past 50 years, the Urban Redevelopment Authority (URA) and fellow agencies have evolved their own brand of solutions to cater to Singapore's needs.

Among these efforts, the most ambitious one is to make Singapore a City in a Garden. The term 'City in a Garden' is an oxymoron. In many places in the world, city development invariably leads to the destruction of greenery. But in Singapore, we believe that nature and biodiversity are primordial to the wellness of the human spirit and the provision of greenery to support nature is central to the planning for quality of life. While there is no hinterland to provide respite from the city, we believe that greenery can be woven deeply into the urban fabric. In essence, the 'City in a Garden' is a vision to immerse the city in pervasive greenery.

The work started in 1963 when Mr. Lee Kuan Yew planted a mempat tree. This led to a sustained effort over the past 50 years that has blossomed into healthy urban ecosystems in Singapore. Today, the city is celebrated for being host to a rich biodiversity. It is not uncommon to see animals that were once thought of as rare or extinct in the urban environment.

Oriental pied hornbills are now frequently sighted in the City. Sea otters are permanent residents in Marina Bay. Lately, we have even been reading about crocodiles on our coast. The best affirmation of the achievements in this area probably took place in 2010, when the Convention of Biodiversity adopted the Singapore Index on Cities' Biodiversity to help cities create plans to enhance biodiversity.

But the city, ultimately, is still an artificial construct. Much of the science we have today has been developed to manage ecosystems in large national parks several times the size of Singapore. In essence, it involves leaving nature alone to thrive. While

T. Schröpfer and S. Menz, *Dense and Green Building Typologies*, SpringerBriefs in Architectural Design and Technology,
https://doi.org/10.1007/978-981-13-0713-3_2

we try to keep aside land for protected nature reserves, this is unfortunately not an option we have for most parts of Singapore.

Having nature in our city, in fact, involves intense human intervention. And to intervene effectively requires the building of a new body of knowledge to help sustain nature in an unnatural setting. In a way, this is what the 'City in a Garden' journey is all about. It is an effort in constant experimentation to push the frontier of knowledge to find new solutions. Whether it is to introduce a new native forest species into the urban environment, convert a concrete canal into a rustic stream, or to grow plants on tall buildings, there are often very few known precedents to guide our work. It often involves months of trials with people from different disciplines to make nature happen in the urban environment.

The Dense and Green Building Typologies symposium is significant because it has brought together some of the best people here in Singapore who have been pushing the frontier of knowledge to enhance the urban ecosystem. Its contributors represent the whole body of knowledge that Singapore has acquired over the past 20 years in this area of work. They all help address an important question: In the dense urban environment, how do we enlist architectural design, buildings and even town typologies to enhance our urban ecosystem?

This is an area where we have done significant experimentation for the past 15 years, starting with modest vertical walls and then trials of roof gardens on public housing plazas and multi-storey car parks.

These efforts cumulated in the introduction of the LUSH policy in 2009 to both incentivise and mandate the provision of skyrise greenery on new developments. As of today, more than 130 hectares of skyrise greenery have been implemented across different development types, ranging from condominiums to shopping centres, offices, hotels and hospitals.

But despite these achievements, there are still significant knowledge gaps in this area of work. For instance, how can buildings and town typologies be better designed to house greenery? How can the quality of greenery be improved to enhance biodiversity? How can it be designed to bring down the cost to ensure long-term sustainability? How can the community be better involved? There is still much work to do to answer some of these questions and contribute to formulating better plans and policies to advance the 'City in a Garden' vision.

Mr. Ng Lang was appointed the Chief Executive Officer (CEO) of the Urban Redevelopment Authority of Singapore (URA) in August 2010. Before this, Lang was CEO of the National Parks Board of Singapore for 5 years, where he played a key role in implementing major green infrastructural programmes to achieve the 'City in a Garden' vision. This includes the development of new parks and the park connector network, streetscape greenery masterplans, skyrise greenery, the expansion of the Singapore Botanic Gardens and the new Gardens by the Bay. He also championed community outreach efforts to enhance public appreciation and participation in Singapore's greening and biodiversity conservation programmes. Prior to his National Parks appointment, he served in various capacities in the Singapore Public Service, including the Singapore Foreign Service and the public healthcare sector. Lang is currently a board member of the Singapore Tourism Board and the Jurong Town Corporation.

Chapter 3
Creating Liveable Density Through a Synthesis of Planning, Design and Greenery

Cheong Koon Hean

What is the Housing and Development Board (HDB) doing regarding biophysical design? About 6–7 years ago, HDB had to triple its housing programme to meet the demand for public housing. While the huge programme was a big challenge, it gave HDB opportunities to build a new generation of public housing. In 2011, HDB developed a roadmap to develop better designed, community-centric, sustainable and smarter towns. As HDB houses more than 80% of the population, it was important for HDB to develop its towns in a sustainable manner as it would contribute to the sustainability goals of Singapore as a whole.

Punggol was HDB's first eco-town. To guide its development, HDB came up with its Sustainability Framework. This framework set out 10 key desirable outcomes for environmental, social and economic sustainability (Fig. 3.1).

This included very detailed strategies and initiatives for greenery and biodiversity as a key outcome. The specific KPIs at town level in this framework were aligned with the national Singapore Sustainability Blueprint. Following that, HDB also developed the HDB Biophilic Town Framework to guide the enhancement of the existing natural assets to achieve a greater sense of place as well as an enhanced quality of life and well-being for residents. This framework elaborates on the strategies and initiatives for 'Enhanced Greenery and Biodiversity', one of the 10 desired outcomes under the broader HDB Sustainable Development Framework.

Traditionally, HDB develops its towns based on a checkerboard pattern, juxtaposing the areas with high and low density to mitigate the perception of high density. The town centre is surrounded by neighbourhoods which have around 4,000–6,000 units each.

Overlaid onto these elements are three tiers of greenery: town parks, neighbourhood parks and green buffers. An example of an older HDB town planned based on the checkerboard principle is Tampines.

In recent years, HDB has also layered on a 'linear green' element onto the existing HDB hierarchy of parks. For instance, in the new area called Tampines North, we have incorporated a boulevard park which can be enjoyed by the surrounding housing. Extending from the Boulevard Park, a 'green tapestry' of linear greens link up to

T. Schröpfer and S. Menz, *Dense and Green Building Typologies*, SpringerBriefs in Architectural Design and Technology,
https://doi.org/10.1007/978-981-13-0713-3_3

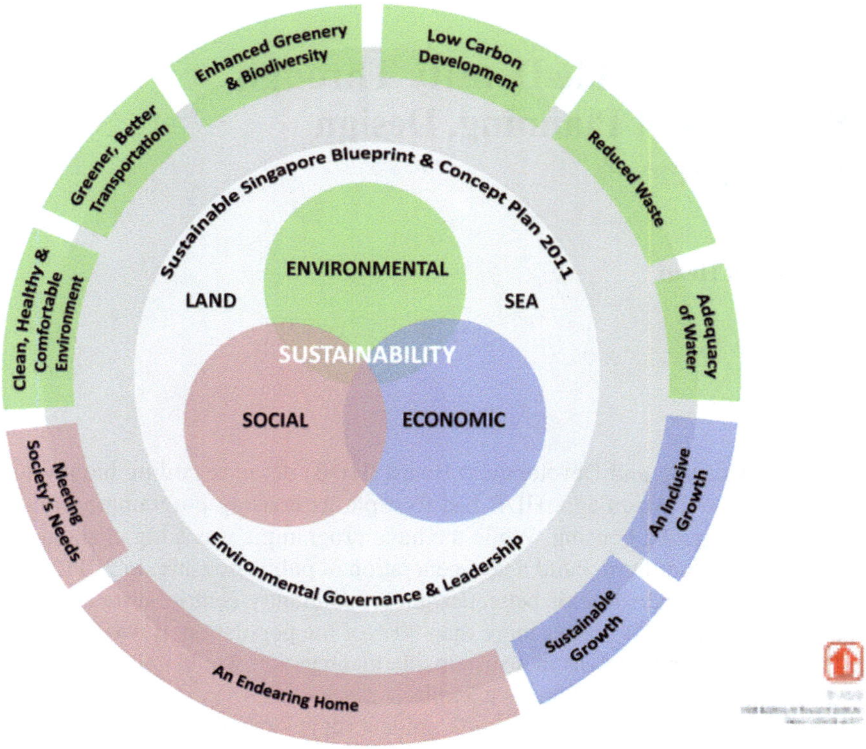

Fig. 3.1 HDB's sustainability framework; *Credit to* Housing and Development Board

smaller local parks which serve as 'green living rooms' located right at the doorsteps of the residents (Fig. 3.2). Together, these provide lush and seamless connectivity from homes to parks.

For Punggol, HDB made adjustments to the traditional spatial structure of the town. The size of neighbourhoods was shrunk from 4,000–6,000 units to about 1,300–2,000 units to provide neighbourhoods with a more intimate feel. In Punggol, common greens and schools were grouped with each neighbourhood to create more accessible greenery. The common green is located next to a school field to give the perception of a larger open space. Where possible, the school field was made available for public use after school hours.

Besides typical town parks, neighbourhood parks and common greens, there are four more layers of greenery in Punggol: green connectors, precinct landscapes, rooftop gardens and skyrise greenery. For instance, the Matilda District in Punggol features inter-precinct greenery—comprising common greens and green connectors and intra-precinct greenery which comprises precinct landscapes and rooftop gardens. Greenery is also provided on raised roof decks as part of the new building typology and these are linked up to expand the sense of space.

Fig. 3.2 Green tapestry; *Credit to* Housing and Development Board

In addition to greenery, we also introduced a 'blue' element to the development of Punggol. Two rivers, the Serangoon and the Punggol River, were dammed up to form two freshwater reservoirs. Instead of a functional pipeline to connect the two reservoirs, we created a river and named it the Punggol Waterway. A functional element, therefore, serves as a recreational element as well.

The developments on either side of Punggol Waterway were sensitively designed to optimise the river as a pleasant living environment. For example, Waterway Terraces is a well-designed development where the building terraces down towards the river, with its landscaped gardens fully integrated with the river promenade.

Oasis Terrace is a mixed-use neighbourhood centre development, which has been sensitively designed to relate well to the Punggol Waterway. This project is currently under construction and should be finished by the end of the year. The neighbourhood centre is directly linked to the adjacent LRT station. A civic plaza, which is a covered public space overlooks the waterway and is specially designed to encourage community activities.

Punggol Northshore is another example of how we integrate new ideas of urban ecosystems into town planning. The planning of this particular precinct emphasizes four aspects of our biophilic landscape design: air quality management, ecological balance, stormwater management, and urban heat island effect mitigation. For air quality management, trees are strategically planted to direct and enhance the wind flow along footpaths. Clusters of trees and shrubs that are effective in air purification are planted at the end of such footpaths to serve as 'air refresh zones'. Ecological

Fig. 3.3 Pinnacle @ Duxton; *Credit to* Housing and Development Board

balance is achieved by increasing the diversity of greenery layers, creating habitat zones such as dragonfly ponds, butterfly gardens, and bird sanctuaries. For stormwater management, district-wide network bioswales and rain gardens are planned to enhance water cleansing, to regulate and to filter stormwater runoffs. To mitigate the Urban Heat Island effect, trees that are high in carbon sequestration rate are selected. The design of the Punggol District Landscape Masterplan was also drawn up through layered mapping and analyses of the site conditions. Through Computational Fluid Dynamic (CFD) studies, trees will be strategically planted to direct and enhance the wind flow along footpaths.

The Pinnacle@Duxton features multiple layers of greenery. Two sky gardens were created for this high-density development as recreational spaces for the residents. The environmental deck above the car park was landscaped and co-located with facilities such as a basketball court and playground for recreation (Fig. 3.3).

HDB also greens many of the roofs of its buildings such as those on top of multi-storey carparks. We also invented and patented a prefab extensive greening system for many older buildings with structures that cannot take additional loading from heavy soil for planting. The advantage of this system is that it is lightweight and requires low maintenance as plants only need to be watered every three weeks.

HDB also adopts water-sensitive design principles in its developments. For example, in collaboration with the Public Utilities Board (PUB), Waterway Ridges was a pilot project, where HDB adopted Water-Sensitive Urban Design (WSUD) for a large precinct for stormwater management. Landscape features here have achieved additional sustainable water management and treatment and at the same time have created interesting spaces and water elements for the community to enjoy. The landform has been shaped very carefully. Twenty-one rain gardens, four vegetated swales, and two gravel swales have then been incorporated to clean the rainwater collected and to lower peak flows into the freshwater reservoir downstream from Punggol Waterway.

In order to help clean water and encourage greater biodiversity, HDB has also designed a floating wetland modular planter system, which was introduced into the Punggol Waterway. This system, which works very effectively, won an ASEAN engineering award.

HDB carries out research with many parties and stakeholders on urban greenery. We are studying the development of a biodiversity index for residential towns to measure the ecological integrity and biodiversity levels of any HDB town. HDB is also working with the National University of Singapore, the Chinese University of Hong Kong, the URA and the National Parks Board (NParks) to research and further develop the Biophilic Town Framework so that it can be used for the planning and design of HDB townships.

Although Singapore is a dense city, we believe that through the careful synthesis of good planning, design and urban greenery, we can create a highly liveable city. Such an integrated approach and the introduction of pervasive and innovative forms of greenery throughout Singapore will take us yet another step closer to fulfilling Singapore's vision of being a 'City in a Garden'.

Dr. Cheong Koon Hean is the CEO of the Housing and Development Board (HDB) overseeing the development and management of some 1 million public housing flats in 26 towns/estates. Dr Cheong was also the CEO of the Urban Redevelopment Authority from 2004 to 2010, in charge of strategic land use planning, conservation of built heritage and the real estate market. She played a key role in the planning and development of major growth areas, such as Marina Bay as well as the Sino-Singapore Tianjin Eco City. She is also a nominating committee member of the Lee Kuan Yew World City Prize. Dr Cheong is on the Boards of the HDB, the National University of Singapore, the Civil Service College as well as the International Federation for Housing and Planning. Active internationally, she served on various expert panels and was a member of the World Economic Forum's Real Estate and Sustainability Global Agenda Council. She is also the Tan Swan Beng Endowed Professor in Nanyang Technological University. She has been conferred several awards, including the Meritorious Service Medal for outstanding public service, the Convocation Medal for Professional Excellence (Newcastle) and the 2011 IWF Woman Who Make A Difference Award (Washington). In 2016, she became the first Asian to be conferred both the Urban Land Institute's JC Nichols Prize for Urban Visionaries and the CTBUH's Lynn S Beedle Lifetime Achievement Award.

Chapter 4
Building a City in Nature

Khoo Teng Chye

Today, Singapore is often discussed in the context of liveability and density but four or five decades ago, it was a completely different story. There were about one million people living in slums. Each one of the low-rise wonderful shop houses that you can see today used to accommodate about 100 people, living in three shifts. For example, in the morning shift, one person would go to work while another would come back to sleep. Singapore was overcrowded, the Singapore River was an open sewer and the City had droughts, floods, crimes and diseases. But its population was less than two million. Today it has become a much more liveable city with a population of about 5.6 million. It means that the population actually tripled and yet the City's liveability improved.

There are all sorts of rankings on liveability. Most of them rank cities like Vancouver, Sydney and Melbourne among the most liveable ones though most of these cities have a rather low density. There are very few cities that have high density and good liveability. That is what Singapore has been able to achieve (Fig. 4.1).

The work of the Centre for Liveable Cities (CLC) is to figure out what has brought Singapore that far and whether it is possible to extract from it some useful principles and then going forward, whether or not those principles still apply as Singapore moves into the future. Singapore is an island and its land area will always remain constrained. However, its population must grow, whether it is to 6.9 million or 7, 8, 9 or 10 million, it has to grow to be able to compete economically. In this context, the question is whether we can achieve liveability as density increases even further.

Achieving liveability is about the quality of life, having a sustainable environment and maintaining a competitive economy, all in a balanced way. In a situation with very little land, no water, no energy, very few resources and a lot of people, this is challenging. Looking at what Singapore has been doing in various areas of endeavour in the urban sphere, whether it is in urban planning, transport, housing or greening and blueing, it has taken what is called an 'urban systems approach', which is about integrated planning, execution and governance (Fig. 4.2).

When people refer to high density in cities worldwide, they tend to show tall buildings and how the urban environment looks like a concrete jungle. But when

T. Schröpfer and S. Menz, *Dense and Green Building Typologies*, SpringerBriefs in Architectural Design and Technology,
https://doi.org/10.1007/978-981-13-0713-3_4

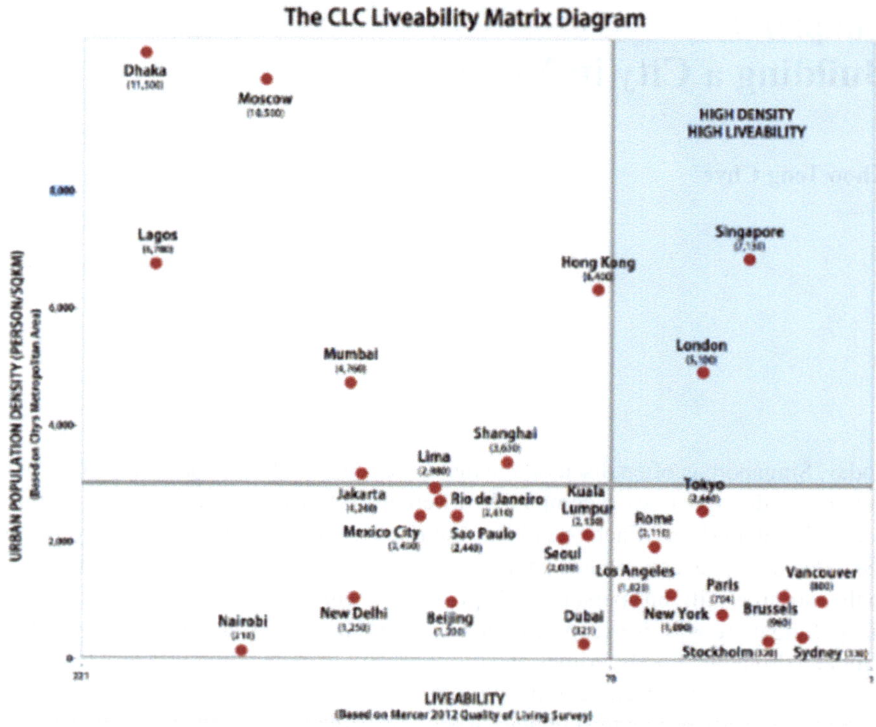

Fig. 4.1 Liveability matrix; *Credit to* Centre for Liveable Cities

visitors walk down Orchard Road, they often say Singapore looks like a city full of trees. Also, very often the images of Singapore show high rises but the environment is actually very green. That is one of the important ingredients for why Singapore is more liveable compared to other cities. Beyond the greenery, with the Active, Beautiful, Clean Waters (ABC Waters) programme started in 2006–2007, Singapore now aims to be a city of gardens and water. And then beyond this, can Singapore also be a city in nature? (Figure 4.3).

How did Singapore become green? It was not about planting trees indiscriminately. It was about making Singapore green in tandem with urban planning and development. This is what is called the 'urban systems approach', having a holistic, comprehensive vision and then being able to translate that into appropriate plans, policies and projects.

Three elements sum up the urban systems approach taken to assure that greenery is pervasive in Singapore. First, in Singapore, there is a hierarchy of greenery. It is not just parks everywhere. We have national parks such as the Botanic Gardens and the Gardens by the Bay, regional parks such as the West Coast Park, the East Coast Park and the Bishan-Ang Mo Kio Park, town parks and within each town we create

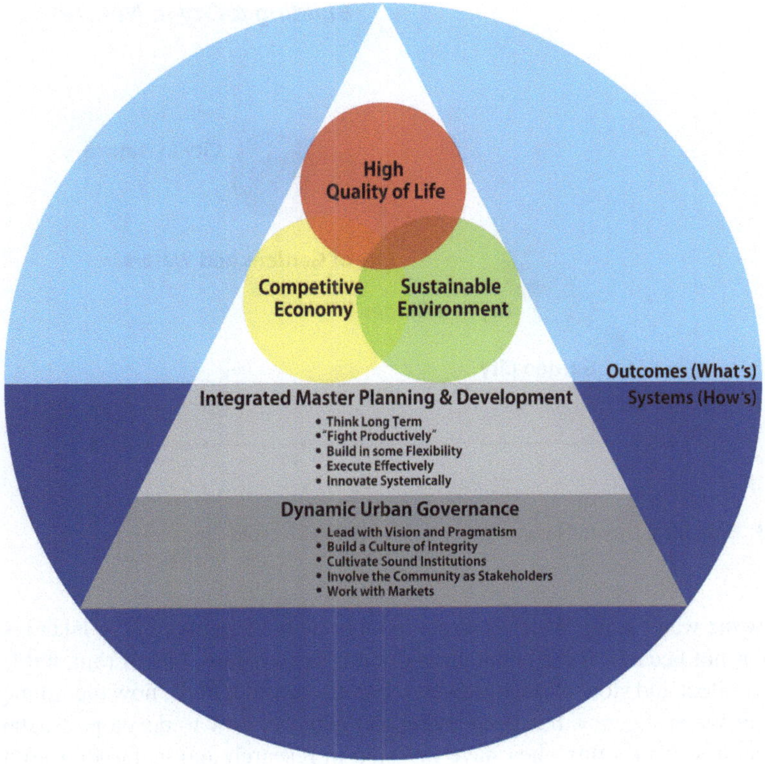

Fig. 4.2 Singapore Liveability Framework; *Credit to* Centre for Liveable Cities

neighbourhood parks and community parks. Hence, there is a planned hierarchy of greenery inserted into the urban matrix, the towns and the buildings.

Second, the idea that Singapore wants to have prolific amounts of greenery. This was articulated in the early days of Singapore by the Garden City Action Committee, putting Singapore under a canopy of trees. But how to achieve that? By overlaying greenery on the various networks that the city is made up of. The most obvious one is the road network. How to make sure that all the roads are lined with trees, what type of trees, the spacing and the standards? Even open parking lots, every three parking lots must have a tree. In terms of drainage, what happens on the side of the storm water drains? Now there is a park connector system. These were standards created in the early days to make Singapore green.

And then, third, all the land in between the buildings should be secured for green buffers and beyond that, greenery on the buildings themselves, whether it is the rooftop or skyrise greenery. These three elements have been implemented systematically, allowing Singapore to go from 35% of greenery in 1986 to almost 50% in 2010 while the population has also doubled during that period. This is the green story.

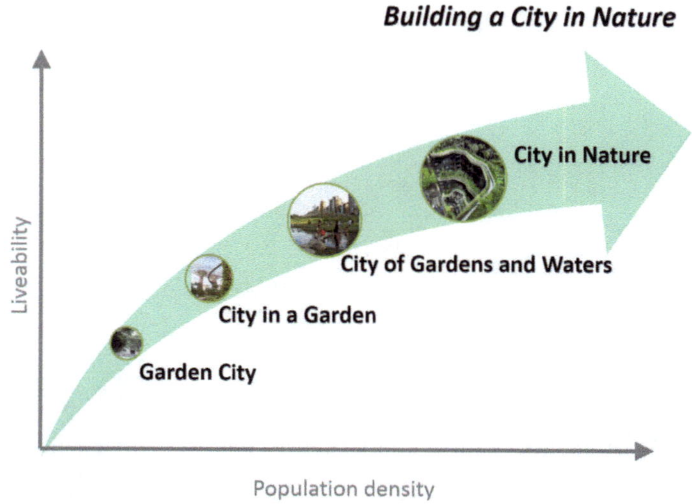

Fig. 4.3 Building a City in Nature; *Credit to* Centre for Liveable Cities

Now the water story. Water is an essential issue for Singapore. The Island is short of water, not because there is not enough rain; there is rain, lots of rain, but limited land to collect and store all the rainwater. For a long time, PUB, now the Singapore's National Water Agency, has been trying to figure out how to develop a sustainable water supply. To do this, they have invested in research and technology to unlock water solutions for Singapore—the recycling of water to produce NEWater and the desalination of water using membrane technology. But one of the most important things that PUB has done is harvesting most of the water from the sky: two-thirds of Singapore is now a water catchment that feeds into a total of 17 reservoirs.

One idea behind the ABC Waters programme has been to get the individuals living in a water catchment to subscribe to the idea that water is a precious resource to them. Beyond that, from an urban planning point of view, it has also been to show the value of water, not just as a resource but also as an environmental and urban asset. When looking at the blue map of Singapore, PUB asked how to turn its 17 reservoirs and its major drains and canals into something more natural, more landscaped, to transform concrete drains, canals and reservoirs into vibrant, beautiful streams, rivers and lakes. It is important to note that the genesis of the ABC Waters programme was actually started by the URA in the late 1980s, when the Waterbodies Design Panel did a number of very interesting projects. Sungei Api Api, for example, which looks like a river with mangroves but it is actually a monsoon drain. Similarly, the Bukit Panjang storm water pond (later renamed Pang Sua Pond) is a very utilitarian concrete structure but was naturally landscaped. Unfortunately, in 2003, these projects had come to an end. There was a lack of systemic approach to sustain and institutionalise them and that is why the ABC Waters programme was started using the systems approach.

Fig. 4.4 ABC programme, Kallang River—BEFORE; *Credit to* PUB, Singapore's National Water Agency

Under the ABC Waters programme, PUB divided Singapore into three catchments through a master plan. Looking at the blue map, it asked where were the opportunities for PUB to create projects to naturalise the water bodies? The master plan identified about a hundred of them and has systematically gotten Government approval to execute the projects in phases.

Very often, ABC Waters projects raise property value. In fact, when Mr. Lee Kuan Yew visited an exhibition of the ABC Waters programme in 2007, the first thing he said was that this is a great programme because it will have the support of the people. Because instead of an ugly drain that flows beside their property, if it turns this into something beautiful, their property value will go up and they will appreciate you for that.

And indeed, that is what has happened in Bishan Park (Figs. 4.4 and 4.5). When PUB made a press announcement about what Bishan was going to look like, the developers immediately took pictures to put them in their showroom galleries and increased their property prices.

Beyond just beautifying waterways, what CLC is now trying to urge PUB and URA to do is to go beyond just looking at the water bodies. We want them to look at the way water flows and think about the land catchment in which it flows and how to do more with the water, how to integrate water into the urban planning process.

Fig. 4.5 ABC programme, Kallang River—AFTER; *Credit to* PUB, Singapore's National Water Agency

Bishan-Ang Mo Kio Park is an example of that. PUB could have just beautified the concrete drain but instead naturalised and blended it with the park.

The parkland is also used as a floodplain. Now instead of a park and a separate drain, there is a very big park with a meandering river fulfilling both functions in the midst of the high-density high-rise towns of Bishan and Ang Mo Kio.

Another example is the Punggol Waterway, which links the Serangoon Reservoir and the Punggol Reservoir. As PUB water engineers always try to make sure that the reservoirs are connected, the idea was to install a pipe in between them. But former Minister Mah Bow Tan asked, why a pipe? Why not opening up that pipe and why not create a waterway? Punggol Waterway is thus another example of thinking of the new opportunities water presents when planning a new town. One last example is the Jurong Town Corporation (JTC) trying to create a new eco-industrial park typology at the fringe of Nanyang Technological University. While designing it, they looked at the topography and, instead of just creating a conventional drainage system, they asked how could they create a wetland? Hence, in the CleanTech Industrial Park there is now a constructed wetland.

There are many ways to integrate water with greenery in a more natural way and to make it part of the urban fabric to improve people's environments and there are lessons that can be learnt from how NParks has developed the green network. NParks is very ambitious in that respect. It has people who understand biodiversity and really see Singapore as a natural ecosystem. There are four biodiversity core areas: the central core which includes the Central Catchment and Bukit Timah Nature Reserve, the western core includes the Sungei Buloh Wetland Reserve, the northern core includes Pulau Ubin and Coney Island and the southern core includes the entire stretch of the Southern Ridges. Rather than leaving them as four cores, they have thought about how to connect them with nature ways, buffer zones and nature parks. Even single projects at a microscale can create opportunities to be part of a broader urban ecosystem.

CLC has made a few recommendations. For instance, beyond parks and water bodies, for the concept plan review, there should be a 'City in Nature' plan. Second recommendation, beyond ABC Waters or water-sensitive urban design, why not further encourage nature-sensitive urban design? These are tremendous opportunities for research but this also comes with many challenges. People will be asking, why even develop a city in nature? Is it for health, resilience or aesthetic reasons? There is a need to understand these aspects to be able to justify this to the Ministry of Finance. Many new projects are currently coming out: HDB is making Tengah a forest town and new opportunities might arise in Pulau Tekong, where a polder method will be used to reclaim land. That land is not going to be used for the next 20–30 years, except by the Ministry of Defence. Hence, Pulau Tekong could be an opportunity for piloting and testing ideas. In the Jurong Lake district as well, the second central business district, the URA is planning to introduce a new waterway to bring the lake closer to the development and to create a lakeside village.

These are all tremendous opportunities for research and Singapore does have the potential to become the first city in nature. For CLC, this is not just a wish but something that we must make sure does happen.

Mr. Khoo Teng Chye is currently the Executive Director of the Centre for Liveable Cities, Ministry of National Development (MND). He was formerly the Chief Executive of PUB, Singapore's National Water Agency (2003 to 2011), President and Chief Executive Officer of Mapletree Investments and Managing Director (Special Projects) of Temasek Holdings (2002 to 2003), Chief Executive Officer/Group President of PSA Corporation (1996 to 2002) and Chief Executive Officer/Chief Planner at the Urban Redevelopment Authority, URA (1992 to 1996). He sits on the boards of the Tropical Marine Science Institute of National University of Singapore (NUS), the GDF Suez's (Engie) Urban Strategy and Innovation Council, the National University Health Pte Ltd (NUHS), the Temasek Foundation Ecosperity and the Temasek Foundation Connects. He serves as a member of the Development Projects Advisory Panel (DPAP) set up by Ministry of Finance (MOF). He is also the Chairman of the Urban Land Institute (ULI), Singapore, Chairman of the Ministry of Manpower, Singapore's Workplace Safety and Health (WSH) Institute Governing Board, a Senior Fellow of the URA Academy and is a member of the Advisory Committee to the Chief Minister for the formation of the New Capital for the state of Andhra Pradesh, India. Mr. Khoo graduated with First Class Honours in Civil Engineering from Monash University, Australia. A President-cum-Colombo Plan Scholar, he also holds a Master of Science in Construction Engineering and a Master of Business Administration degree from the National University of Singapore. He is a Fellow of the Institution of Engineers, Singapore. He attended the Advanced Management Programme at the Harvard Business School.

Chapter 5
From Garden City to City in a Garden and Beyond

Yeo Meng Tong

When Singapore became independent on 9 August 1965, it was an island nation without hinterland and resources. In order to develop the City despite the lack of a domestic market and natural resources, the Government needed investors. Singapore abided by the simple principle for survival in order to compete with its neighbours—to be more efficient and better organised than others in the region. One of the ideas that the Prime Minister at that time, Mr. Lee Kuan Yew, had to achieve this was to line the passageway from the airport to Singapore's CBD and the Istana with trees and shrubs, so that investors would be convinced of the capabilities and resilience of Singaporeans:

> Visiting CEOs used to call on me before making investment decisions. I thought the best way to convince them was to ensure that the roads from the airport to their hotel and to my office were neat and spruce, lined with shrubs and trees. When they drove into Istana domain, they would see right in the heart of the city a green oasis… Without a word being said, they would know that Singaporeans were competent, disciplined and reliable…
>
> – Excerpt from Lee Kuan Yew's memoir, From *Third World to First World, 1965–2000: The Singapore Story*, page 81.

When asked, which one was Singapore's most important building or monument, famed architect Norman Foster replied:

> Why do you need important buildings or monuments? When I get off at the airport, the drive from the airport to downtown is your most important building and monument.

A government unit was created to execute and manage these planting plans and this unit eventually became the National Parks Board (NParks). It was known as the Parks and Recreation Division under the Government's Public Works Department (PWD) in 1973 and then it became the Parks and Recreation Department under the Ministry of National Development in 1975. In 1995, it formally became NParks. On 16 June 1963, the Greening Singapore Campaign was honoured by Mr. Lee Kuan Yew with a tree-planting ceremony, a practice that is still carried out today by leaders and members of local communities.

T. Schröpfer and S. Menz, *Dense and Green Building Typologies*, SpringerBriefs in Architectural Design and Technology,
https://doi.org/10.1007/978-981-13-0713-3_5

Fig. 5.1 East Coast Parkway; *Credit to* National Parks Board

In 1975, the East Coast Park was created on 180 hectares of reclaimed land along the sea. Mr. Lee felt that the best way to enjoy the seaside was when it was built as a park, where it is open to everyone, whether rich or poor, 24 h a day. This reclaimed area is now covered with trees and its many rain tree-lined pathways are often the subject of iconic pictures of Singapore's greenery (Fig. 5.1). Since then, NParks has created many regional and neighbourhood parks.

Another focus for NParks has been to increase land optimisation. The first successful project to do so was the Raffles Place Park. From 1900 to 1964, the area was used as a typical British marketplace and then as a carpark. The area was then developed to house an underground carpark with a park commissioned on the area above it. Today, the underground carpark has been replaced by the Raffles Place MRT station. The park still exists today, surrounded by tall commercial buildings in the heart of Singapore's Central Business District and it is used by professionals for relaxation away from the office.

To green the streetscape and create a Garden City, NParks introduced guidelines for promoting and maintaining the greenery of Singapore. These guidelines included the conservation of trees and mandatory landscaping components for newly developed projects. Currently, NParks has about 1.4 million trees planted in its parks and about 400,000 of them are rain trees. Fortunately, trees grow fast in the tropics and they can reach full maturity and great heights within 20–30 years. NParks also selects plant species based on their desired characteristics, such as hardiness and fast growth. In addition to green coverage, NParks also selects plants that produce aesthetically pleasing, fragrant and unique flowers. This is to have a variety of blooming colours throughout the year, giving the impression of multiple seasons in Singapore. NParks also incorporates unique planting strategies to integrate vegetation into infrastructure in a visually appealing way. For example, the road safety barriers in Singapore are often lined with small hedges to give the impression that highways are lined with shrubs without compromising on safety.

In the 2000s, NParks entered a new stage of development with a new focus on introducing biodiversity back to the City. It aimed to create streetscapes with different levels and varieties of plants to attract wildlife and to increase biodiversity.

The rich greenery in Singapore also attracts foreign companies to set up their offices here. In 2004, Lucas Films of the famed Star Wars series set up its first overseas headquarters in One-North, in western Singapore. An article published in Newsweek in 2006 that discussed the US animation studio's decision for its first foreign studio, stated that 'Singapore had won the contest hands down, thanks in part to an attribute entirely off the spreadsheet: trees'.

Mr. Lee was very influential for NParks' development, Singapore's greening movement and the transition of Singapore's planning paradigm from a 'Garden City' to a 'City in a Garden'. Three months after the death of Mr. Lee, his son and current Prime Minister, Mr. Lee Hsien Loong, was interviewed by the Times Magazine. Of his late father, the younger Mr. Lee said, very rarely did he (Mr. Lee Kuan Yew) assert a strong view and one of which was the greening of the country. As Singapore progresses into the future, NParks will continue to fulfill Mr. Lee's legacy in interesting and innovative ways. In the Ministry of National Development's Master Plan 2013, NParks was given a key performance indicator that at least 85% of Singapore's residents must live within 400 m of a park by 2030. However, NParks is convinced that it can further reduce this number to 0 m, and that residents can experience a park from their doorsteps. To do so, NParks thinks beyond the boundaries set by master plans to come up with innovative solutions.

The distribution of Singapore's greenery from past to present can be thought of as a network of nodes and connections with multiple dimensions. In the middle of the network is a large area of greenery, the Central Nature Reserve. It is surrounded by green nodes of varying sizes representing the regional and neighbourhood parks and green streetscape connectors linking some nodes.

To increase connectivity, NParks developed the Park Connectors. This is another example of an NParks land optimisation project. In the 1980s, many people were using the areas in and around canals in Singapore for exercise. However, this was a dangerous activity due to their sudden high water levels when it rained heavily. There was a strip of land about 4–6 m wide that was used by the Drainage Department (today PUB) for maintenance twice a year. In land-scarce Singapore, the Government felt these strips of land could be better utilised. As a result, Singapore's first park connector, the Kallang Park Connector, was opened in 1990 by Minister S. Dhanabalan. The first generation park connectors were simple utilitarian designs with a running track and minimal landscaping along their perimeters (Fig. 5.2).

Fig. 5.2 First-generation park connectors and land optimisation; *Credit to* National Parks Board

Fig. 5.3 Second-generation of park connectors; *Credit to* National Parks Board

Today's park connectors are very different. These second-generation park connectors are the result of collaborations between NParks and other Government agencies such as PUB and HDB (Fig. 5.3). Moving forward, NParks is thinking about using its experience from the Southern Ridges elevated walkway at Kent Ridge to roll out a new, third generation park connector typology. This elevated concept could be applied to the canal between Bukit Timah Road and Dunearn Road to avoid demolishing the surrounding vegetation and to retain tall trees, creating a linear skypark along a riverine forest in the City.

To further increase the intensity of greenery in the City, another typology in the Singapore greenery network was created over the past 20–30 years: 'Skyrise Greenery'. An example of this is WOHA's Oasia Downtown, a 27-storey commercial building with plants growing on its facade. This created a giant 'tree' in the skyline of Singapore. Another example is Dr. Ken Yeang's Solaris, a 16-storey commercial building located in One-North. What makes this building unique is that it has a 1.5 km-long continuous pathway of greenery that links the ground floor to two rooftop gardens. This could be thought of as a vertical park connector and as an idea for fourth-generation park connectors. Instead of running vertical marathons in cramped fire escape stairwells, one could utilise green-lined slopes such as the one in Solaris.

One can push the boundaries of imagination even further to create more connections and nodes in the green network for people living and working in Singapore by taking inspiration from nature. Sinkholes are natural phenomena found in many parts of the world. There is the idea that a sinkhole could be used to create more vertical space for living and working, with a vertical park in the centre and vertical water harvesting practices in place. Residential areas could be built above the ground level, while retail and office space could be built below ground level. In the centre, a vertical park could be created so residents can live next to a park. When it rains, the rainwater would irrigate the layers of vegetation as it would flow downwards through the vertical park. The excess rainwater would travel downward into water storage tanks at the very bottom. With some treatment, the water storage area could also become an underground reservoir. For this idea to be realized, it would need the cooperation of many public and private agencies such as the Urban Redevelopment Authority (URA), the Building Construction Authority (BCA), JTC Corporation, the Economic Development Board (EDB), PUB, HDB, to name just a few.

How can we go even further to create more vegetation in Singapore's green network? In the past, NParks has been able to create parks in mangrove, beach and marine areas, but it is possible to go one step further by developing greenery on the sea. The Sea Tree idea in the Netherlands can be used as an inspiration. The project is a floating structure that consists of many layers of habitat for flora and fauna both above and below water. When completed, it will be a haven for wildlife. This idea could be adapted to build floating desalination plants offshore so Singapore's land space could be more effectively used. Using the approximately 690 sq km of sea area selectively can almost double the total area available, giving 1430 sq km, some of which can be set aside for innovative park and green development projects such as these.

Greenery is the landmark of Singapore. Everywhere one goes, one sees greenery. The parks are used daily. Singapore is not just a 'Garden City', but a 'City in a Garden'. A new era in NParks has begun. It has adapted its theme to reflect Singapore's movement to become a 'City in a Garden', continuing to explore interesting innovations for the future.

Mr. Yeo Meng Tong received a degree in Landscape Architecture from the Berlin Technische Fachhochschule, Germany and Master of Science in Recreation Management from Loughborough University, United Kingdom. Mr. Yeo has been a staff member of National Parks Board since 1988. He has been responsible for the planning, design and construction of parks, green spaces and park connectors for more than 25 years. Major projects include Park Connectors, Pasir Ris Park, Yishun Park, the redevelopment of East Coast Park, nature parks and Singapore Botanic Gardens. He has done the development and management of the rooftop garden for Singapore Pavilion in Shanghai World EXPO (2010). He is also providing planning and design consultancy/advisory services to local governments in other countries, including the Sino-Singapore Friendship Garden design advisory to China Tianjin Ecocity local government and the Park Connector planning advisory to Colombo local government in Sri-Lanka. He is now doing design research on the future of urban greenery, which includes therapeutic gardens and space optimization.

Chapter 6
Punggol Waterway Terraces, Singapore

Manuel Der Hagopian

The utopic vision of Glen Small's 'Biomorphic Biosphere Megastructure' can be seen as one of the key inspirations behind our project Punggol Waterway Terraces. This megastructure overcomes the boundary between landscape and architecture, between inside and outside. Similarly, the vision behind Punggol Waterway Terraces was the imagination of a concept close to utopia. The challenge was to build high density without creating a tower block solution.

10 years of experience in Southeast Asia suggest that clients often default to a generic solution for density in the form of a tower. Punggol Waterway Terraces is based on a concept that proposes an alternative. Through the use of explorational study models, the volume evolved from typical building blocks to 'in-between' blocks and resulted in a hybrid form (Fig. 6.1).

This conceptual model (Fig. 6.2) provided hints at the earlier stage about how landscape can be integrated with the massing and how the two can combine verticality and horizontality. It also demonstrated the potential of an organic geometry to embrace the volume of open space. Even though it was a naïve diagram, it still had many of the design ideas that eventually got materialized. The porosity of the mass, the open space that embraced what became the units of the project, the 'jungle courtyard' and the stepping of the massing that allows green to extend onto the top of the buildings are some of these ideas.

In Punggol Waterway Terraces, there is no hard boundary between landscape and architecture. In that sense, the project looks more like a part of the landscape than architecture. The shape of the massing is the result of the voids in between. The voids are the 'jungle courtyards' that bring the waterway landscape from the outside to the inside of the development (Fig. 6.3). It is actually the landscape that shapes the entire project and makes it contextual.

The key elements of the layout are the hexagonal modules called 'cells'. Each cell is formed by 'wings' surrounding one of the 'jungle courtyards'. The hexagon as a shape enabled us to avoid the need for a rectangular geometry for the plan. It resembles the biological or the organic cell geometry that is characterized by 120-degree angles. At the same time, it is an open field of exploration for the architect.

T. Schröpfer and S. Menz, *Dense and Green Building Typologies*, SpringerBriefs in Architectural Design and Technology,
https://doi.org/10.1007/978-981-13-0713-3_6

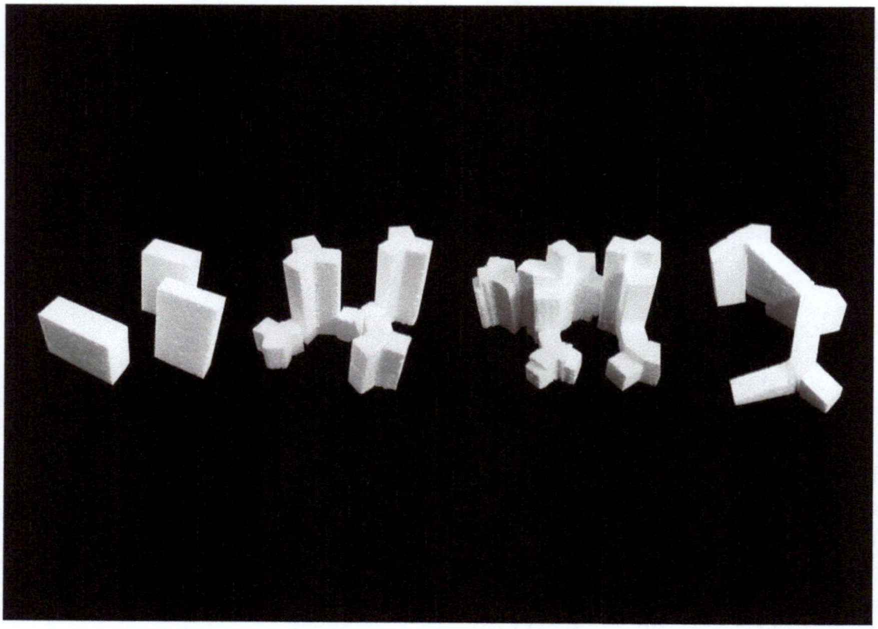

Fig. 6.1 Exploration models; *Credit to* G8A

Fig. 6.2 Design concept; *Credit to* G8A

As a module, it can be adapted to a lattice or repeated as a network of cells and it can potentially grow in all directions (Fig. 6.4).

The 'jungle courtyards' are open towards the waterway to achieve both spatial and visual continuity. The configuration of typical wings around the courtyards creates different degrees of the enclosure for looking outside. These courtyards are also among the best places in the project for residents to enjoy shade both at the semi-basement and ground levels (Fig. 6.5).

Fig. 6.3 First-stage competition model; *Credit to* G8A

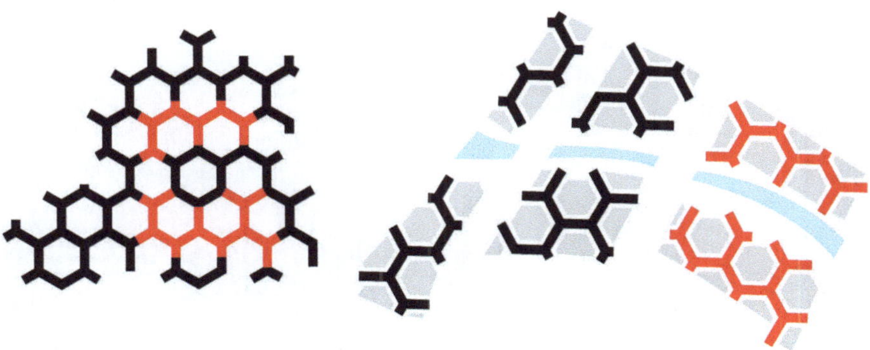

Fig. 6.4 Concept diagram for hexagonal layout; *Credit to* G8A

The second stage competition model (Fig. 6.6) also represents the design ideas that have been realized. First, one can see the cascading of the massing like an erosion of a hill from the top-down to the waterway. The proportion is dramatic enough to create a real massing rather than just a roof effect. Then, from the canal to the top, one can also see the green seamlessly integrating with the project at different levels and scales. At the top, roof terraces provide views to the canal, while the planters proposed at this stage are still visible.

Fig. 6.5 Jungle Courtyard; *Credit to* Patrick Bingham-Hall

Fig. 6.6 First-stage competition model; *Credit to* G8A

Once the hexagonal shape was fixed as the basic module for the project, the scale and the height of the cascades were adjusted. Some early photos of the building document the cascading effect from the top to the lower levels and the green on the lower terraces within the community space. The overall effect of the configu-

Fig. 6.7 View from the Punggol Waterway Park; *Credit to* Patrick Bingham-Hall

ration facilitates different perceptions of the building for the viewer. Depending on one's viewpoint, the building looks like a new tower, a mid-rise or a courtyard block (Fig. 6.7) but it is essentially more of a horizontal 'organism' than a harsh vertical structure and it allows for the generation of more community attachment and sentiment.

The cores of the building are interlinked throughout the entire block, which allows people to go from one unit to another. This also allows for access to the roof terraces. This linked-corridor layout is also advantageous when one is approaching the project from the adjacent Light Rail Transit (LRT) and Metro Rapid Transit (MRT) stations.

Each floor of the project features double-loaded corridors with three-sided lobbies. The porosity of the mass helps to facilitate cross-ventilation (Fig. 6.8). From indoor to outdoor, porosity also allows views to the courtyards. It also makes this massive building seem like a dentelle and appear light and fragile from the outside.

Inspired by the vernacular idea of the roof, the nature-inspired curves of the facade (Fig. 6.9) were designed to protect the building from rain and sun. Curves in nature were also the major reference for creating irregularity with a precast facade element which had to be used as this is an HDB project with a limited budget. This detail of the facade also functions as a balcony which maintains the strong visual relationship between residential units and the lush surrounding nature.

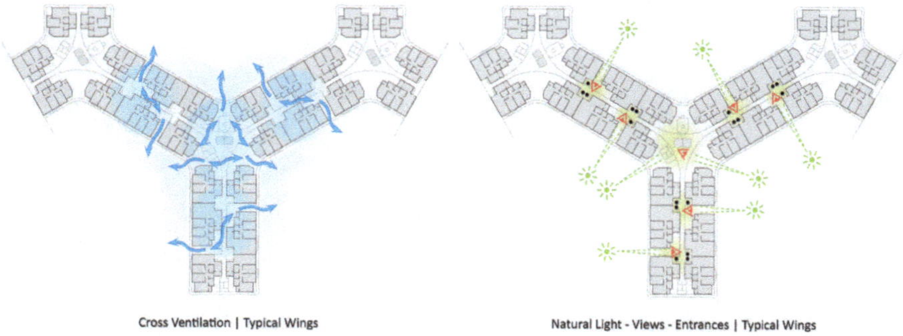

Fig. 6.8 Cross ventilation and natural lighting of a typical wing; *Credit to* G8A

Fig. 6.9 Curve facade and porosity of the building mass; *Credit to* Patrick Bingham-Hall

In conclusion, the Punggol Waterway Terraces project is an attempt to bring liveability to a dense urban environment through connections to nature. The key idea is that nature itself can be a major source of inspiration in the design process.

Manuel Der Hagopian graduated from the Geneva Institute of Architecture, Switzerland (IAUG) and the Belleville School of Architecture, Paris (ENSAPB). He is one of G8A's co-founders and partner in charge of operations in the Singapore office, as well as co-management of offices in Geneva, Hanoi and Ho Chi Minh City with co-founder and partner, Grégoire Du Pasquier.

His main responsibilities include design direction, conceptual design and project management. Notable projects under the direction of Manuel include Punggol Waterway Terraces (Singapore), North Tampines Residents (Singapore), The Alps Residents (Singapore), FPT Soft Village 3 and 4 (Hanoi, Vietnam). His notable projects in Switzerland include numerous residential projects such as White Ships (Bulle), Bellevue project and Twin houses (Geneva). Through G8A's think tank, aptly named 8+, he is currently pursuing his strong interest in the consequences of a rapidly growing population's impact on residential housing in South East Asia. His study focuses on the correlation between evolutionary paradigms of Singapore and Hong Kong. Since 2015, he has been a guest professor at the Singapore University of Technology and Design (SUTD), where he aims to strengthen the transfer of architectural and design knowledge between Singapore and Switzerland.

Chapter 7
Prototypology and the Twenty-First Century

Richard Hassell

The Garden City vision was developed in the Twentieth Century and had at its heart the destruction of industrialization and the desire to keep nature in our cities. However, the scale of cities in Asia is so large that they cannot integrate nature or function by copying what other cities did in the Twentieth Century. That approach simply uses too much space for its large populations. While Singapore is a positive outcome of the application of many Twentieth Century planning principles, often when these are applied elsewhere in Asia, they result in dispiriting cities. They lead to problems rather than to solutions. There is a need to start looking at urbanism in highly dense Asian cities as 'macro-architecture'. To provide liveable solutions, cities must be conceived in three dimensions rather than two and as an interconnected whole at all levels, from large systems down to small components. Another strategy to provide liveable solutions to high density is to layer cities to provide multiple ground planes.

As buildings within Asian cities get larger and increase in density, they also get three-dimensional and the spaces within them also start to get very big. The need then arises to start thinking of buildings themselves as sub-cities and considering how people perceive them. Buildings cannot just have corridors and functional spaces. They need public spaces and a type of urban form that makes it easy to navigate through them. When designing a building as a city, it is also important to carefully ensure that everything within them operates simultaneously to achieve multiple goals, including meeting infrastructure requirements as well as providing leisure opportunities and different community and public spaces.

How can these large buildings be comfortable and efficient? When creating high-density buildings in the tropics, a designer's first instinct is often to create very large volumes, which are then treated as indoor spaces, mimicking strategies that are implemented in temperate areas, to make them more comfortable. However, lower energy solutions can be used by considering factors that affect air movement and breathing such as building orientation. When implementing greenery in buildings, it is important to understand that plants also need light and air, need to be looked after and change over time . Testing, innovation and good strategies are needed to

T. Schröpfer and S. Menz, *Dense and Green Building Typologies*, SpringerBriefs in Architectural Design and Technology,
https://doi.org/10.1007/978-981-13-0713-3_7

implement this idea and to maintain its appearance. Over time, plants may die or they may have to be replaced, which is often a concern for building owners. However, on the ground, this is something that is accepted all the time and this is perhaps something to learn from.

In Singapore, the climate is tropical, making the design of skygardens easier than in many other locations. However, Singapore has been important as a prototype for greenery in architecture for its surrounding region. Through Singapore's example and WOHA's in-house research, designing strategies for much more hostile environments, such as those that have water shortages, places with typhoons, cyclones, etc. have been developed. Although some environments are more 'hostile', sometimes there can be a higher tolerance for badly maintained and failing buildings. However, this phenomenon may also be linked to a lower state of governance and maintenance of buildings.

One example of such a project in Singapore is Skyville@Dawson (Fig. 7.1). This is an interesting example because many people had doubts about whether as a typology

Fig. 7.1 Skyville at Dawson, Singapore; *Credit to* Patrick Bingham-Hall

Fig. 7.2 Oasia Hotel Downtown, Singapore; *Credit to* Patrick Bingham-Hall

for affordable housing it could be applied elsewhere in the City. The ideas proposed were seen as very expensive and only meant for luxury projects. However, Skyville was built with a tightly controlled budget.

Oasia Downtown (Fig. 7.2) is another interesting project in this context as it features extensive greening and cladding. In terms of Leaf Area and Green Plot Ratio, it stands at over 1,000%. One could imagine inserting this typology across an existing cityscape and having one building compensate for 10 other buildings around it which have no greening, for example conservation buildings and buildings that are difficult to retrofit.

In the Community Town Hub project in Tampines (Fig. 7.3), the sports club and sports arena are on the roof and this creates a giant tropical umbrella for the rest of the development, enabling a large area of land to be released on the ground.

The scheme for a university in Dhaka, Bangladesh (Fig. 7.4) has a water remedial ground plain. The building is flooded with greenery. This could work as a catalyst to show an alternative direction for developments in this city.

Fig. 7.3 Tampines Town Hub, Singapore; *Credit to* WOHA

Greenery in a city can be in the form of a cladding material like creepers or as gardens as part of individual developments but also can be found in larger aggregates such as parks, interconnected buildings or a whole master-planned development. Larger parks, public ones, in particular, are a great way to get people invested in the idea of Skyrise Greenery. At this larger scale, thinking of buildings as topographies can generate much more interesting environments to be in rather than huge buildings with big slabs of greenery on top. One example of this is the Kampung Admiralty project, which has just been completed (Fig. 7.5). It features a mix of components that are integrated three-dimensionally with a park on top.

Fig. 7.4 BRAC University, Dhaka; *Credit to* WOHA

Another example is one of WOHA's projects in Penang that creates a significant real estate in the sky by the careful placement of high-rise towers and sky parks in between buildings. Another project in Mumbai has a 4-hectare car park on 11 levels, 4 underground and 7 above (Fig. 7.6). This creates an enormous amount of built space in the city. Luckily, it has a very irregular site boundary, which made it conducive to an interesting topography and landscape, turning it into an artificial hill.

There are so many apartments within the development (it has a population of 20,000–30,000 people) that the parks are large enough to function as a public space even though the development is private. This idea of topographic architecture as a way of handling large unattractive masses completely changes the relationship people have with architecture and allows it to be more humane.

Fig. 7.5 Kampung Admiralty, Singapore; *Credit to* Darren Soh

Another option for creating interconnected greenery at a larger scale is by connecting individual projects through skybridges and parks. Back in 2001, in our scheme for Duxton Plain (Fig. 7.7), we proposed that every 5 floors should have a public space as well as streets and gardens.

A whole city of these buildings would create opportunities for horizontal connections between them. To implement this, planners must first identify several three-dimensional datum lines where buildings must be connected. An urban area would have cobbles or launch pads or structures to support bridges and connections between buildings. WOHA's projects have been constructed to be compatible, should this system be implemented. Newton Suites and The Met in Bangkok are two examples. In The Met (Fig. 7.8), there are two full levels at the 28th and 47th storey where it would be possible to connect with an adjacent building.

Fig. 7.6 New Cuffe Parade, Mumbai; *Credit to* WOHA

One landscape-scale project designed by WOHA is the Tengah Forest (Fig. 7.9). This project was designed to be fully self-sufficient in terms of energy, vegetables and water needs. It was also designed with the idea in mind that Singapore can improve further in terms of producing food, water and energy in one area of the island, which also has great benefits in terms of security.

Fig. 7.7 Duxton Plain, Singapore; *Credit to* WOHA

How do we make interconnected projects and skyrise greenery landscapes more commonplace? One issue to address is the negative loop between planners and architects. Planners feel like they should be very conservative in what they propose as components to their master plan in case developers say it is too difficult to achieve. This entrains an old-fashioned Twentieth Century building like a podium and tower typology or a standard residential block into a master plan. Therefore, when architects want to innovate and break out of the more traditional way of designing, they are limited by out-of-date Twentieth Century objects. To break out of this cycle, it

Fig. 7.8 The Met, Bangkok; *Credit to* Patrick Bingham-Hall

Fig. 7.9 Tengah Self-Sufficient City, Singapore; *Credit to* WOHA

is very important to look at projects as prototypes for a city of the future. If these prototypes exist, planners could conceive of much more radical cities and they could populate them with more interesting and useful buildings.

Richard Hassell is the Co-Founding Director of WOHA. He graduated from the University of Western Australia in 1989 and was awarded a Master of Architecture degree from RMIT University, Melbourne in 2002. He has lectured at many universities and served as an Adjunct Professor at the University of Western Australia. Hassell gave a lecture at ETH Zürich, under the Future Cities' Lab European-South East Asian Architectural Dialogue in November 2016 together with Wong Mun Summ. He has mentored students at the National University of Singapore in the Embedded Studio in Practice programme, anchored the University's Master of Science in Integrated Sustainable Design Masterclass since 2011 and conducted a design studio at the Singapore University of Technology and Design in 2016. WOHA exhibited in Mexico City at the Museo de la Ciudad de Mexico from March – April 2017 as part of the 2017 Mextropoli Festival. They also participated at the 2016 Venice Architectural Biennale. In conjunction with their installation in Venice, WOHA has launched a new book titled "Garden City Mega City", which shares strategies for the burgeoning mega cities of the tropical belt.

Chapter 8
RGB: Red, Blue, Green as a Model for Living Environments

Franklin Po

Certain important principles have always guided the design ethos of Tierra Design and PODesign. These include the integration of the allied design disciplines—Landscape Architecture (as many of our professionals come from this field), Architecture, (because some of these same professionals were trained as architects) and Urban Design (because the combination of landscape and architecture fully illustrates how people live and use dynamic spaces). Using this ethos, Tierra Design and PODesign have worked towards liveability and love-ability or what they consider as living in a biophilic city where humans have an inherent connection with nature. For designers, the challenge then is to ensure that their designs are for living in a city that works for people.

The journey to greening started modestly in the year 2000, when Tierra designed the landscape for a small condominium called 'The Loft' (Fig. 8.1) in collaboration with W Architects on a small plot of land on Nassim Hill. This project is considered one of Tierra's first designs of vertical green. A 50 m-long architectural wall that Tierra usurped and brought into its design was incorporated with green landscape strips over a modular granite clad wall. This was done to hide an irrigation system. As one of the first vertical green expressions, this became an integral part of the architecture. Modular PVC cells were applied onto the granite walls and grass was planted into this system. At this time in Singapore (2001), there were no other similar systems. Nobody had attempted any kind of vertical planting design and maintenance was an issue because organic coco-peat instead of an inorganic planting media was used which in time must be replaced.

158 Cecil Street (Fig. 8.2) was an invitation from architect Kelvin Kan to create an intervention for a blighted interior atrium space of seven storeys. The challenge was to make the atrium in this building more attractive and rentable. Layers of vertical greenery were added that rose from the first to the seventh floor inside the atrium. Insufficient natural light from only south-facing windows meant that plants would not thrive. The solution was to augment the space with artificial lighting to ensure that the plants would have enough light for photosynthesis and chlorophyll production. A potted planting system was designed and hung from new structural

T. Schröpfer and S. Menz, *Dense and Green Building Typologies*, SpringerBriefs in Architectural Design and Technology,
https://doi.org/10.1007/978-981-13-0713-3_8

Fig. 8.1 The Loft, Singapore; *Credit to* Tierra Design

elements which allowed for a 400 mm maintenance space behind the potted plants. The vertical greening intervention completely transformed the interior space and the offices which now look out into the garden atrium were soon all rented. For some time, it was home to Google Singapore until the company moved into a much larger facility.

Another recently completed project is the landscaping for the National University of Singapore's Natural History Museum (Fig. 8.3), a building designed by W

Fig. 8.2 158 Cecil street, Singapore; *Credit to* Tierra Design

Fig. 8.3 Natural History Museum, Singapore; *Credit to* Tierra Design

Architects. This project's green intervention is different because the design utilised planting as part of the architecture and is easily accessible from the occupied floors.

Fig. 8.4 Changi airport terminal 3, Singapore; *Credit to* Tierra Design

In the year 2000, Tierra won a competition to facilitate the interior planting of Changi Airport Terminal 3 (Fig. 8.4), which took a total of 8 years to build. After analysing the spatial layout, the design took its influence from the concept of a greenhouse, utilising all the skylights above it. There are three layers to this project: the entrance layer crossing the bridges, the transitional area between the land side and air side and the air side lounges along the tarmac.

Another project is a private condominium development by Wingtai called 'The Tembusu' (Fig. 8.5). This project was designed to complement the principles of PUB's Active Beautiful and Clean (ABC) Waters programme. It has a vegetative swale, soak-away planting beds and rain gardens. It is a successful project even though there were some compromises, something many projects unfortunately suffer from because of budget constraints.

A much larger urban scale mixed-use project called IAPM was designed and implemented in Shanghai (Fig. 8.6). Tierra was asked to design the streetscape as well as a small garden at the corner of the site. However, its design and urban site context had hidden opportunities as all the buildings in the southern portion were single-storey retail shops located next to an MTR terminal building. Going beyond the design brief, the architectural configuration gave Tierra a chance to green the rising site and also the architecture both vertically and horizontally. Planting was installed over the vertical sides and on the top of the building structure.

Tierra and POD are determined to work on future landscape solutions. Central to our vision is to design with biophilic principles in mind. By 'biophilic' we mean a family of organisms that are in a friendly relationship. We must always consider that

Fig. 8.5 The Tembusu, Singapore; *Credit to* Tierra Design

we only have one planet and one city that we call home. There are many projects today which offer biophilic strategies that we can learn from and adapt and apply their principles in many other places. For example, creating soak-away planting beds for drainage systems alongside roads and incorporating urban farming projects to create a community (kampong) spirit within a garden setting.

Another issue to consider is the ageing population of Singapore, which will be significantly larger by 2030. Tierra wants to help our seniors to age well in a safe and friendly environment. The ultimate goal is to help keeping our elderly population fit and engaged so as to slow down the effects of ageing.

As for our design approach, developments can be measured by a few metrics. A green plot ratio (for landscape design), a blue plot ratio (for water management), a yellow plot ratio (for community inclusiveness) and a red plot ratio (for road efficiency). As building designs have advanced, these metrics have undergone progressive changes. Firstly, the blue plot ratio was improved by adopting PUB's ABC Waters incentive programme and the green plot ratio was improved by utilising the Green Mark scheme encouraging 100% site area replacement with more greenery on buildings. Developers started trying to make mega-developments more accessible by adding multiple skygardens and more green elements to buildings to further enhance the green plot ratio. More recently, clusters of these mega-developments have been conceived. These match density with even more liveability with buildings connecting at higher levels to create a truly biophilic design between buildings for people (Fig. 8.7). Within these developments, people feel that they can walk out and enjoy green spaces anytime, reducing their stress levels when they are in nature. As

Fig. 8.6 IAPM, Shanghai; *Credit to* Tierra Design

buildings have evolved in this way, more spaces have been progressively developed for walking, biking or using personal mobility devices, which, in turn, result in fewer cars on the main roads, but still allowing for taxis and buses. Meanwhile, opportunities for community engagement in the form of urban farming have increased, leading to more community interaction and a higher yellow plot ratio.

In the Jurong Lake District Plan project, Tierra addressed challenges to integrate buildings with strategies to lower the road plot ratio (Fig. 8.8). Within the project, we replicated the sense of climbing a mountain of green in different buildings. Overall, we drew our inspiration from the kampong (community) spirit and put in place the hardware, the 'heartware' and the software by making the whole development open, connected and interactive. People want a healthy life and ultimately happiness. Tierra tries to pursue this through public and private partnerships and Singapore is the right place to explore and to take the lead on the greening of the city.

When designing the Jurong Lake District project, Tierra also looked at how to implement interconnected biophilic infrastructure from the scale of individual land parcels all the way to the urban scale. The project was proposed as a biophilic canal city that provides environmental, economic and social resiliency, lower road plot ratios, greening of different layers with connections ranging in heights from underground to sky terraces and skywalks, while utilising various modes of transport and even personal mobility devices (Fig. 8.9). Tierra is determined to create such

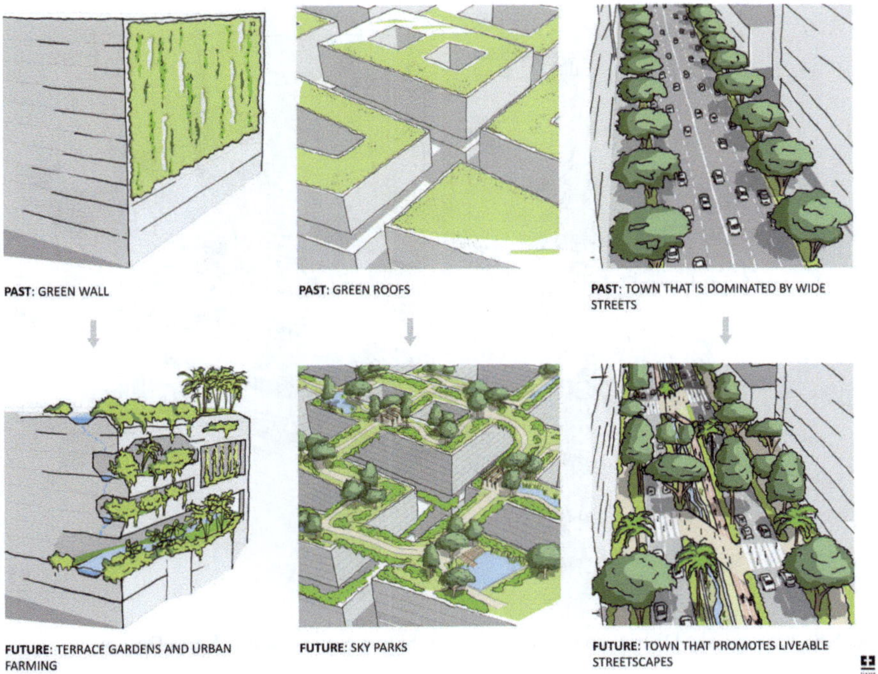

Fig. 8.7 Evolution of building design in Singapore; *Credit to* Tierra Design

environments as they are inevitably our future. With such ideas, there is an attempt to integrate the community and the city to create a biophilic city. Other fundamental questions that Tierra explored are possibilities to increase the green plot ratio to have food within the city itself, within its architecture. This can be achieved by placing sky farms and productive gardens at various locations. The project also aims to infuse water throughout the whole infrastructure, taking water from the lake and distributing it through the new business district.

Another project completed by Tierra several years ago was the landscape consultancy for the Marina Barrage (Fig. 8.10). The goal was to conceptualise the building to be more than just a pump station but also a public space.

More recently, Tierra has completed the concept design for Singapore's fourth desalination plant, a building that is 107 m wide by 220 m long and at its apex 13 m high. Sinking the building 7 m into the ground enabled the plant to look less monumental. The design allows water from the freshwater reservoir and the sea to flow into the desalination plant without pumps as part of the desalination process. Another key design strategy was to capture all the rain that falls on the building's surface and channelling it into a retention area through a vegetative swale to cleanse the water. While the result of this process is not potable water, it is sufficiently filtered to be used for water features designed around the entire building. During heavy rainfall, the system uses a bypass pipe to channel the water directly into a

Fig. 8.8 Jurong Lake District conceptual drawing, Singapore; *Credit to* Tierra Design

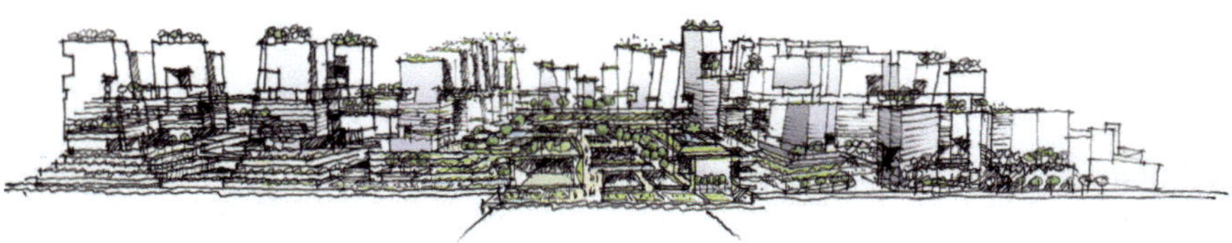

Fig. 8.9 Jurong Lake District conceptual drawing, Singapore; *Credit to* Tierra Design

culvert. These rainwater collection strategies could be utilised in future buildings by collecting all the water from the building itself, channelling it through planters, filtering it and then harvesting it for use. Once the water has gone through all these systems, it gets displayed as water features and cascading walls of water and finally falls into a retention pond where it is collected for use and recycled again.

Fig. 8.10 Marina barrage, Singapore; *Credit to* Tierra Design

Mr. Franklin Po's journey of many years in the design discipline had its roots at the University of California, Irvine. These early years witnessed the beginnings of his interest in art and design. Artists Vija Celmins, Craig Kaufman, Robert Irwin, Ed Moses and Larry Bell were all influences. Celmins introduced him to the work of Van Gogh and Max Ernst, to Dada, to Surrealism, to Man Ray, to The Bauhaus and to the architects Walter Gropius and Mies van der Rohe. Enrolling in architecture classes at the California State Polytechnic University he met Richard Neutra, Buckminster Fuller and Raymond Kappe (Founder of SCI-Arc). A Welton Beckett fellowship enabled Franklin to attend the UCLA Graduate School of Architecture and Planning. Graduating in 1973 with a Masters Degree in Architecture with Design Honors, he nonetheless was taught a tough lesson by architect Yoshio Taniguchi that would serve him well during his 20 years of practice in Los Angeles. Tanaguchi's sharp words continue to remind him to examine his work critically: "…SO WHAT!" Franklin continues to guide Tierra's focus as a strong proponent of integrative design and landscape urbanism. In 2015, Franklin won the Singapore President*s Design Award Designer of the Year.

Chapter 9
How Blue–Green Infrastructure Can Create Liveable Cities and Address Climate Change

Ryan Shubin

The Ramboll Studio Dreiseitl approach to blue–green infrastructure is characterized by a deep integration of water and landscape. We have an almost obsessive curiosity about water and its interactions with land—how it performs, how it carves out spaces, how it works in the natural system, its internal behaviour, its meandering nature, why it works the way it does, what it looks like, not only as an art piece or inspiration for beauty but also as a logic in a landscape (Fig. 9.1). Through this obsession, Ramboll Studio Dreiseitl is well positioned to understand water within the urban context, where it is often overcontrolled and underutilised.

Often water is considered as just another element of infrastructure, similar to a road or an electrical line. However, when water takes full control of a city, such as during the 2017 super storms in Hong Kong and Houston, it essentially dominates the entire terrain. In the natural system, flexible measures allow for water in the landscape. In the urban system, it is generally underestimated, which results in, for example, the historic floods of Copenhagen in July 2011 and Beijing in July 2012. Both events caused damages in the amount of close to a billion Euros.

These flooding incidents are framed in the context of climate change. While it is difficult to pin any particular storm to climate change, the general consensus from climate research is that extreme weather events are going to become more severe and frequent. It is not exclusively about storms, climate change can lead to ultra-dry spells as well. The urban fabric has to adapt to not only the water logics inherent to natural systems but also the overall effects of climate change on hydrology. Blue–green infrastructure is a key element that Ramboll Studio Dreiseitl often champions to address these problems.

The urban fabric tends to have an abundance of storm water surface-runoff, typically between 70 to 100%. Ramboll Studio Dreiseitl looks towards the behaviour of a more natural system, where only 45% of water is allowed to surface-flow off the site. The remaining storm water is subsequently evaporated or infiltrated into the ground. This results in a more harmonious system where the peak water flow is much lower than in the typical urban condition. As water is a resource, the natural system contains storm water and the effects of runoff are more decentralized. The

T. Schröpfer and S. Menz, *Dense and Green Building Typologies*, SpringerBriefs in Architectural Design and Technology,
https://doi.org/10.1007/978-981-13-0713-3_9

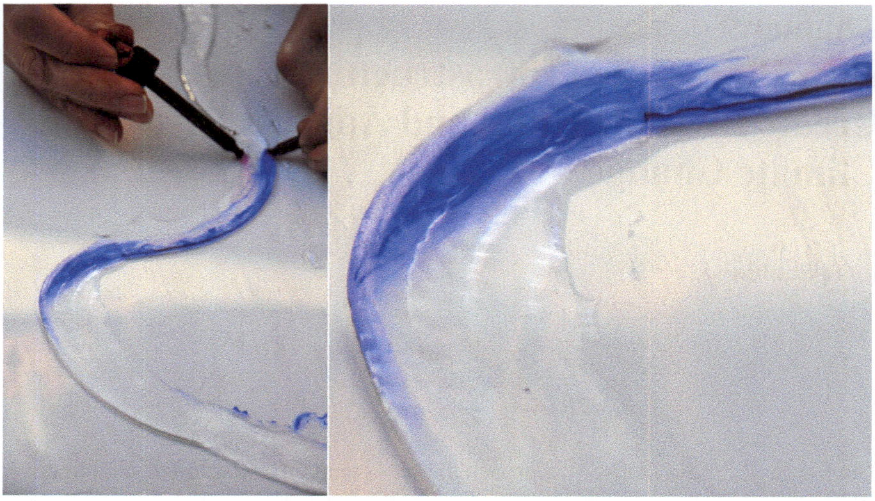

Fig. 9.1 Internal logic of water; *Credit to* Ramboll Studio Dreiseitl

reduction of peak water flow during storm events is one key aspect of Singapore's blue-green infrastructure programme ABC Waters.

The ABC Waters programme looks at how water can become active, beautiful and clean (ABC), thereby making the city more liveable and allowing people to connect safely with water in the urban context. To take advantage of strategies such as ABC Waters, a paradigm shift is required. The traditional sequence of first designing buildings and infrastructure and then adapting the landscape to fit in between needs to be reversed. The inherent values of the natural systems should be first understood before buildings and infrastructure are integrated into the site. This shift in thinking greatly influences the Ramboll Studio Dreiseitl approach.

A case study demonstrating the integration of ABC Waters on top of buildings is Kampong Admiralty, a recently completed Housing and Development Board (HDB) project under the partnership of Ramboll Studio Dreiseitl and WOHA Architects. The project explores how landscape can be integrated into the building, both as an intense system of urban greenery for wildlife and community but also as a storm water management system. The realisation of the project was made possible by a collaborative effort by multiple agencies, with HDB leading the overall project.

Within Kampong Admiralty (Fig. 9.2), located in Singapore's Woodlands district, there is village life and an integration of community. The project is executed through highly sustainable measures and integrates nature and water through biodiversity-sensitive urban design. The emphasis is on giving people access to healthy public living and opportunities to play as well as on creating a vibrant educational experience in outdoor gardens. A beautiful integration of nature-filled planted terraces can be seen in the central courtyard and the food court plaza as well as the playscapes. Water is allowed to flow on the various levels and down into the rain garden.

Fig. 9.2 Urban Village, Kampung Admiralty, Singapore; *Credit to* Ramboll Studio Dreiseitl

To achieve all these diverse objectives on the relatively small 0.9-hectare site with a height limit of 45 m, the stacking of programmes became a necessity. The scheme is based on a 'vertical Kampong' approach, with a people's plaza on the lower level, a medical centre on the mid-level, and a community park with studio apartments for seniors and a child care centre on the upper level. These three distinct strata generate opportunities for diversity of cross-programming and inter-generational bonding through the close proximity of healthcare facilities, social spaces and commercial amenities.

The landscape integrates into the architecture in a series of tiers, allowing for continuous greenery from top to bottom to flourish throughout the building. On the upper stratum, families can come together in open space community gardens and walk through groves of fruit trees. The middle stratum features a rain garden which provides a serene landscape for the medical centre and also serves as a collection area for storm water to percolate down to the harvesting tank (Fig. 9.3). The lower stratum allows for an integration of the Mass Rapid Transit (MRT) station below and features an eco-pond, swales and a cleansing biotope. Overall, the development keeps both natural and urban systems well balanced with a strong integration of terraced ecologies as a frame for an abundance of active community spaces.

The landscape design carefully considers the selection of trees and their strategic planting locations. Local fruit trees, once plentiful in kampongs, were planted as a nostalgic reference to the past for seniors and to garner future interest by the youth. A range of understory planting schemes were adopted to explore a variety of flora compositions to mimic the wild and attract biodiversity such as butterflies and small birds. In addition, a change of seasons was reflected in the lush landscape created

Fig. 9.3 Healing Courtyard, Kampung Admiralty, Singapore; *Credit to* Ramboll Studio Dreiseitl

with specific plants selected for their colour-changing foliage. Efficient storm water harvesting and irrigation systems were developed for the extensive landscaped areas (Fig. 9.4). The tiered architecture provides a larger surface area for collecting storm water runoffs. Taking advantage of the vertical nature of the building, cleansing systems use gravity to channel water through a series of stacked rain gardens which form a continuous storm water cleansing treatment network to purify and harvest rainwater for irrigation. With the integration of ABC Waters design features at Kampung Admiralty, it is important to know how much water can be captured and how much can be used. Based on the average annual rainfall in Singapore, 4.1 million

Fig. 9.4 Vegetative storm water filter, Kampung Admiralty, Singapore; *Credit to* Ramboll Studio Dreiseitl

litres of tap water can be saved each year when storm water runoff is stored in the rainwater harvesting tank and reused for irrigation.

The Punggol Waterway Ridges project is a relevant case study for the integration of the ABC Waters programme between buildings. The incorporation of ABC Waters began in the planning stages through a coordinated effort between HDB and PUB, the national water agency. In typical ABC Waters projects, the design features are retrofitted to suit the existing HDB precincts. The blue–green system at Punggol Waterway Ridges was developed to employ strategies to collect, treat, store and potentially re-use the site's storm water runoff. Here we mimicked nature's hydrological systems to improve storm water quality and lower peak flow runoff to pre-development conditions in addition to creating valuable social and ecological spaces. Storm water passes through several ABC Waters design features to naturally cleanse the rainwater runoff before discharging it into the public waterways. Selected amounts of treated water are channelled to public spaces for engagement and education purposes. The Punggol Waterway Ridges project demonstrates that with a blue–green network in place, approximately 57% of the site's storm water runoff can be naturally treated. These ABC Waters design features were actively engineered to become an integration of both nature and public space in which community life can flourish.

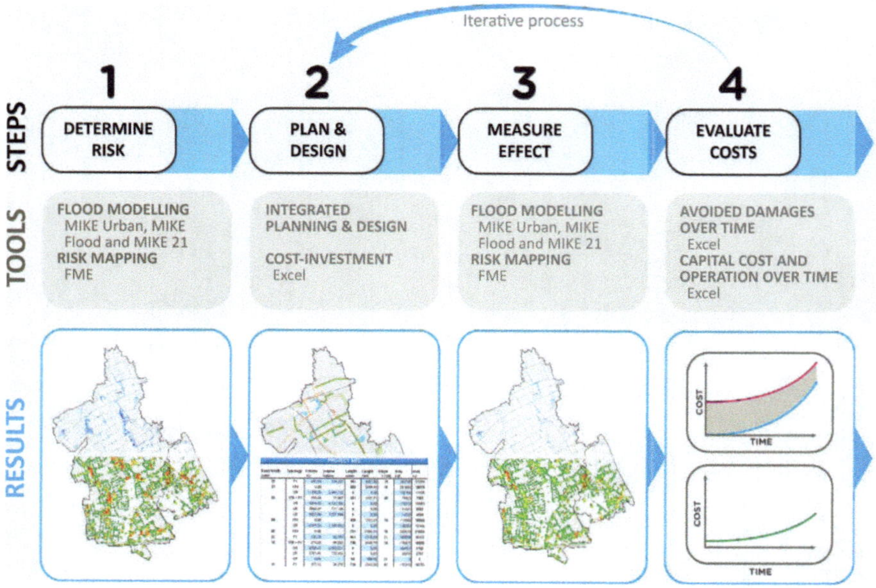

Fig. 9.5 The Cloudburst Resiliency Approach; *Credit to* Ramboll Studio Dreiseitl

Looking forward at how to maximise the value of ABC Waters and blue—green infrastructure, these design features should be understood as dynamic multi-functional spaces with layered programming adapted to the site's full range of climate conditions. For instance, a rain garden or storm retention area can serve as a playfield and educational space during dry days. Swales can create unique restful niches for families to dine, while gently sloped natural waterways can double as playscapes. ABC Waters design features can be designed as sculptural static artworks during non-storm events, then transform into active water cleansing elements during a rain. The future of blue–green infrastructure is to design multidimensional spaces that are adaptable to both climate and social demands and seamlessly integrated into the public realm.

Beyond the building scale, Ramboll Studio Dreiseitl and Ramboll have recently been engaged to explore the integration of blue–green infrastructure across the greater cityscape. Here the focus is on creating a comprehensive cloudburst toolkit in which a municipality can deploy tactically throughout their existing urban environment to adapt to the evolving conditions of climate change (Fig. 9.5).

A prime example of this climate resiliency planning is Ramboll's New York City study commissioned by the NYC Department of Environmental Protection (DEP). Here the aim was to understand if Ramboll's previous Copenhagen approach to storm water resilience planning could be adapted for New York City through a study in the Southeast Queens neighbourhood. First, an examination was conducted of how current floods occur and how flood events in 100 years will likely take place. Through these two data sets, the spatial risks were determined in both the existing and future flood conditions. With this understanding, a master plan was presented of primarily blue–green interventions created to handle the projected 100-year storm levels alongside associated cost estimates for construction and maintenance. The masterplan was developed through a typical design process of considering historic waterways, elevation maps, terrain-based flowlines, along with social and transport infrastructure. The planning process was supplemented by workshops with various New York City agencies through which valuable and relevant social spaces, green spaces and waterways were identified. Subsequently, this information formed the basis for both a direct cost analysis in addition to a cost–benefit analysis.

Within a cloudburst masterplan (Fig. 9.6), design elements in the toolkit include features such as cloudburst streets. Here specific streetscape spaces are prioritised to allow for the capacity of flood detention during heavy storm events while high-value elements, such as adjacent buildings, are less impacted. This requires a variation in design of the existing streets for intentional flooding to occur at centre medians or other localised retention zones. The cloudburst pipe is another element which can be installed below ground for temporary water storage. In addition to elements along the streets, adjacent nature spaces or community gardens may be intentionally depressed to serve as floodable area. Much of the cloudburst toolkit was developed one by one and is based on our more than 30 years of blue–green infrastructure experience. Now, it is assembled as a larger toolkit for citywide climate resilient planning.

Lastly, the cost–benefit analysis shows that grey infrastructure has a higher cost of installation with a slightly lower cost of maintenance when compared to blue–green infrastructure elements. Nevertheless, it is clear that the risks from flooding will be much greater now and in the future if the city chooses not to adopt preventative strategies or infrastructures. Furthermore, the cost–benefit analysis found that despite the initial cost of the blue-green infrastructure elements amounting to USD 330 million, USD 600 million worth of benefits would be gained. In other words, for every dollar spent on blue–green infrastructure, the city would receive 1.8 dollars in societal value. The study has been communicated through an executive summary, which now serves as an instrumental document in the acceleration of climate-resilience planning in New York City.

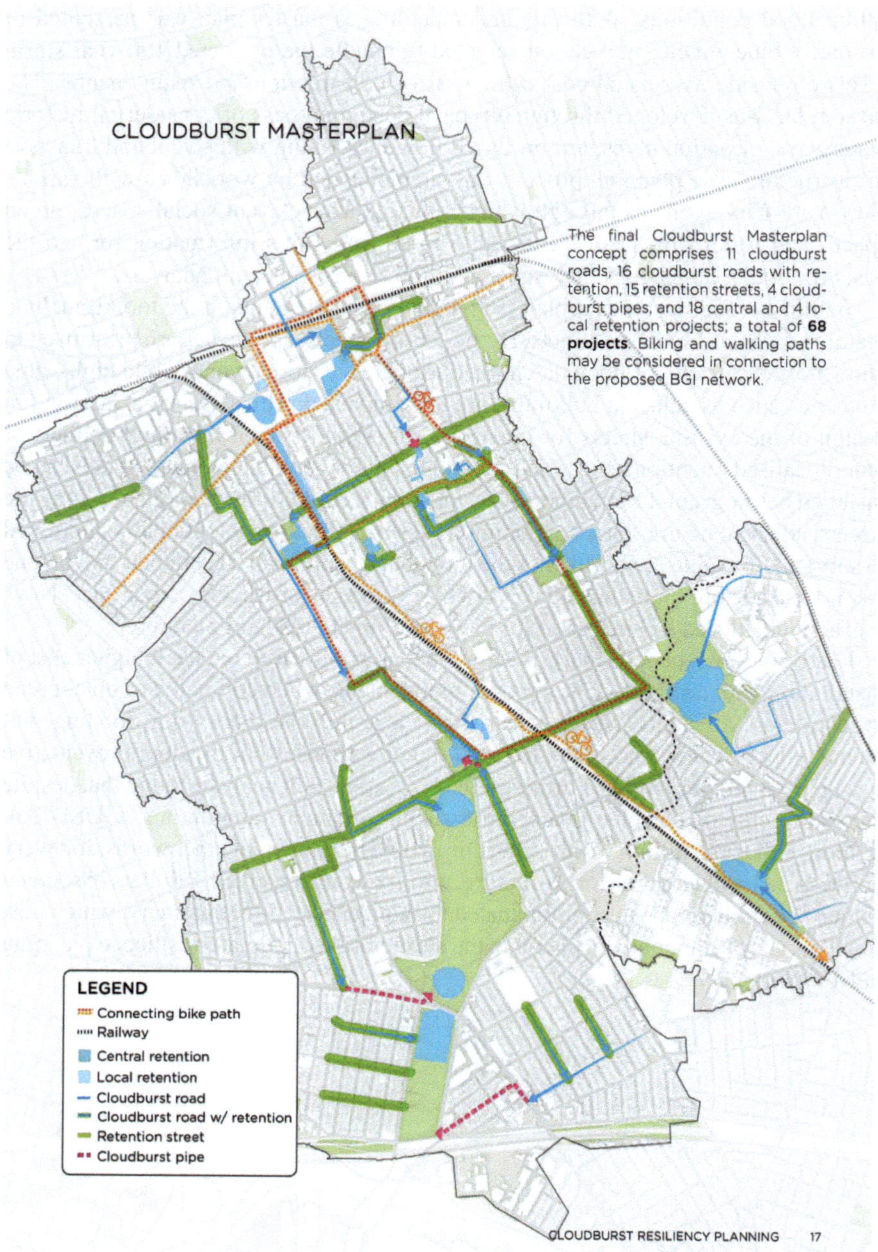

Fig. 9.6 Cloudburst Master Plan of South East Queens, New York City; *Credit to* Ramboll Studio Dreiseitl

Ryan Shubin is Associate and Senior Landscape Architect at Ramboll Studio Dreiseitl (RSD). At RSD, he has led a range of projects from waterfront parks to public show gardens to large scale hydrologic landscape master plans. Ryan has over 11 years of experience in the field of Landscape Architecture. His recent work includes leading the design of the Jurong Lake Garden West, including the construction of this 62-hectare Singapore National Garden located along Jurong Lake; The Friendship Park proposal for the Tianjin Eco-City Riverfront Park Competition; and the ABC Water projects at Kallang and Bukit Batok, two water-sensitive urban landscape designs along the Kallang River and the Bukit Batok canal for Singapore's Public Utility Board. Additional RSD projects include the Jurong Eco Garden, the Jewel Island at Danga Bay, and the National Orchid Garden at the Singapore Botanic Gardens. Ryan's focus lies in the human experience of public urban landscapes designed in relation to greater ecological and hydrological systems. His approach focuses on site materiality, native vegetation and environmental design strategies. Over the course of his career he has been involved in a variety of projects throughout North America, along with various other countries such as the UAE, Chile, China, India, Malaysia and Singapore.

Chapter 10
Chua Chu Kang MKPL's Investigation

Siew Man Kok

Two years ago, MKPL Architects won the competition for the Choa Chu Kang (CCK) section of the Rail Corridor, which extends to 24 km from the north to the south of Singapore. The brief called for a proposal for 3,000 units of affordable housing. The site has varying height constraints with an inspiring context. A barren trail of the Rail Corridor remains on the site bounded by a concrete canal that is 30 m wide and 3 m deep. The Rail Corridor is also disrupted by the Light Rapid Transit (LRT) development as well as a canal. A park connector runs along the side of the canal, opposite to the site.

If the site had been accepted as it is, ignoring the canal and park connector outside its boundary, which is how development is normally conceived, it would have been more of the same. Instead, the proposal boldly broke the competition rules by advocating something that is not practiced today, which is an integrated land use approach in which development simultaneously looks at park design, infrastructure design as well as housing design, all in one go. The proposal rethinks current land use models and re-imagines the way we can live, creating a very efficient yet very liveable environment. The decision was made to reforest the whole Rail Corridor into a 50 m -wide forest. The word 'forest' suggests the element of time—and that makes a difference to a place like Singapore which is so used to instant development. MKPL proposes a development where it takes time for the community to grow, just like it takes time to grow a forest. The proposal imagines a place in that the barren Rail Corridor is reforested and becomes an ecological spine. The concrete canal disappears into the landscape as it is integrated within the park that can be enjoyed by the community.

The whole canal, in parts where it matters, can also be broken up, such as in Bishan Park, except that housing is proposed to be integrated with it—something that carries a lot of risk in the mind of the authorities. At a different section, the canal is decked over to link up with the neighbouring community. The forest will weave through the whole precinct, merging it into the blue and creating two important moments in the site where it connects via a footbridge across the road and LRT as well as to the existing community around it, leading people into the precinct town centre.

T. Schröpfer and S. Menz, *Dense and Green Building Typologies*, SpringerBriefs in Architectural Design and Technology,
https://doi.org/10.1007/978-981-13-0713-3_10

Fig. 10.1 Chua Chu Kang ground-level plan; *Credit to* MKPL Architects

Similarly, the car parks are integrated into the landscape, which are still based on existing HDB-ways of calculating requirements. However, the amount of car parking space in the development could be reduced, or maybe completely removed, as the population's transportation habits change.

The Rail Corridor will become a major pedestrian route, which will be used by visitors all the time. Even now, it is already a very popular cycling route. At the same time, it can become the backbone of the whole community and provide more routes for an even larger community to use the Rail Corridor in the future. More importantly, what MKPL wants is for this community to have a sense of ownership of the whole Rail Corridor. For instance, the process of planting the landscape is designed to be carried out by the residents, giving them ownership of what is being planted. In other words, the community will grow in the way the forest will grow.

The proposal also includes cycling routes. Very light local roads will be networked into the precinct, further connecting the existing and the new CCK communities. The precinct will also be designed to be walkable and elderly-friendly. In discussions with the authorities, we even suggested that the place can do away with roads altogether. But all these elements are just the framework. What is important is the inspirational and aspirational aspect of the whole proposal.

Despite of the existing height constraints, MKPL has created different scales of environments throughout the site, giving each area its own identity (Fig. 10.1). The first element is the Pang Sua Canal that is opened up as a foreground to the 36-storey towers that rise on both sides of the forest. These towers are all connected through the canopy of the forest. The second element comes with the buildings that are literally situated over a flood plain—on rainy days, the ground floor and part of the void deck will be filled with water.

On the Forest Terrace will stand low-rise developments of about seven to eight storeys. There, one can imagine a very simple precinct centre with a market, including amenities like a barber shop, a post office, a library, a child care centre, a senior care centre, all integrated. An 8 m-wide super-bridge designed to rise gently to cross over the LRT, road and canal disruptions of the Rail Corridor will land on the roof of the buildings at the town centre (Fig. 10.2), bringing visitors seamlessly to this part

Fig. 10.2 Village Centre; *Credit to* MKPL Architects

of the development, thereby allowing them the to actively use the Rail Corridor not only for leisure and recreation but for everyday activities as well.

Another aspect is the reforestation of the Rail Corridor. With the proper planning and selection of the right types of vegetation, one can envision joggers mingling with the community, extending into this part of the forest. This will enable the community to have a deeper connection with nature, the temperature, humidity and the sounds and sights associated with a forest.

After the rain, the landscape will become a different place for all to enjoy (Fig. 10.3). Of course, an immediate reaction to this image is the potential danger of kids falling into the water. Nevertheless, this is something that should be revisited. Singapore has now become so risk-averse that it prevents a lot of possibilities of how people can enjoy the Singapore weather and features it creates in this kind of development.

Part of the inspiration for this proposal came from the fact that in a forest, there are different experiences you can get, depending on what height you are on in the canopy. But to provide this experience to the residents, a forest must grow. This is a difficult undertaking, but considering that Singapore has done so much in the past 50 years to green the whole island, it should not be impossible to grow a forest in this area that could reach up to 30 storeys in 30 years. Again, this idea needs careful nurturing with the community and that is where the soft part of the development comes in. It means informing about the benefits that this kind of forest structure can have as well as creating a setting to allow for the community to come in and participate in the development, for example through the community planting trees.

Another challenge is to achieve the two difficult crossings bridging the LRT and the canal. The proposal came up with a super-bridge, which gently rises from a

Fig. 10.3 Wetlands; *Credit to* MKPL Architects

distance away, bringing people up to the canopy of the forest before gently coming down to the town centre (Fig. 10.4). This will give people a new way to experience the Rail Corridor in the future.

Not only integrating greenery in the development was important, the whole development is exploring different building technologies that will address some of the existing problems such as shortage of manpower, the ageing population and future-proofing. Specifically, MKPL decided to use a modular design approach to build all 3,000 units of the community. With the idea that the future home is just a universal space, the size of rooms can be changed to meet the current needs of the residents. Creating a more permeable space between living areas and bedrooms is simple, essentially the use of these areas is only differentiated by furniture. Toilets, kitchens and yards are seen as service modules serving the space. Walls are removable, allowing conversion of the apartments for other uses in the future as the needs arise, such as resident gardens, nursing home facilities and medical centres within the apartment blocks.

The modular system allows for connections between sky gardens in large tower blocks and mid-rise buildings, creating vertical hierarchies of space. The low-rise buildings can be connected at roof level to create more community activity areas such as rooftop farming gardens (Fig. 10.5).

Fig. 10.4 Exterior linear forest; *Credit to* MKPL Architects

Fig. 10.5 Rooftop community farm; *Credit to* MKPL Architects

The modular system designed for this project makes it possible to build cheaply and quickly and at the same time to bring in flexibility that can address future needs. This idea is further developed to create even more connected super-highrise green public spaces, whereby it is possible to create different clusters of community activities in the sky. These are not unbuildable or just dreams, but real solutions that could be implemented in the near future. This can be achieved by the collaboration of different Government agencies so that connections can be made in the sky between different developments, such as those within and surrounding the Rail Corridor.

Mr. Siew Man Kok co-founded and built MKPL Architects into a firm that is highly regarded for its design creativity. Over the last 22 years, the firm garnered several awards including the RIBA Award for Design Excellence for their Kent Vale Faculty Housing project and the most coveted Singapore President's Design Award for the same project. Man Kok was also the winner of the President's Design Award Designer of the Year in 2015. The firm's portfolio includes master planning, residential, mixed use development as well as landscape planning, both locally and regionally. His firm also made a compelling proposal for a housing project along the Singapore Rail Corridor, which won the prize in this international competition. Man Kok's leadership in the design of the various projects within the MKPL Architects ensures that the firm's design ethos and values are not compromised even in the most challenging of circumstances.

Chapter 11
Taking Urban Greening to a Higher Level

Henry Steed

With the new framework of Landscape Replacement Area (LRA) and the Green Mark requirements, implementation of greenery in future developments will be increasingly complicated. This can be observed in early examples of developments under the LRA policy such as Jem and Westgate, where the fragmentation of buildings and the variation of roof levels work towards creating the maximum amount of outdoor space on each of these buildings (Fig. 11.1). Because of this, it is now an opportune moment to look at taking some very pragmatic matters that arise in the movement to green the city to a higher level.

In 1824, Singapore was largely rural with a small town, with most of the island being naturally vegetated. 160 years later, Singapore had become a megalopolis, a huge concrete, steel and asphalt entity only just starting to realise that something had to be done about its environment. It was beginning to turn into a concrete jungle, in spite of the fact that there were huge areas of greenery being installed. In the 1980s, there was hardly a green building in Singapore but that has changed in the past 30 years. Landscapes are now created in the sky and a most influential impetus for this change in urban design is the Landscape Replacement Area. Today, the LRA is essentially about creating environments by replacing the land that was taken for buildings with 60% hard landscapes and 40% soft landscapes on top of the buildings in various forms. 100% replacement is only a practical figure for its first stage. The LRA should not simply be a rule to be complied with. In other words, it is more than just a policy with a number-crunching exercise attached. It is instead a national philosophy which transforms the environments on top of structures into usable, attractive and connected spaces—gardens, spaces and streets—for the benefit of the community and people using the buildings. Now, in 2017, Singapore has gone green, bursting out of the seams with excellent green initiatives and green ideas. However, there is no point if this is all talk and no action—drivers of the green movement need to implement urban greenery and look into how best to do it.

How can we turn Singapore's green dream into a reality? As the LRA represents the developer's drive to create quality environments for communities, creativity should be unleashed to optimise the 100% replacement, in regards to maximising benefits

T. Schröpfer and S. Menz, *Dense and Green Building Typologies*, SpringerBriefs in Architectural Design and Technology,
https://doi.org/10.1007/978-981-13-0713-3_11

Fig. 11.1 JEM West, Singapore; *Credit to* ICN Design International

(or even to 150%) with costs being only a moderating constraint. For example, if grass needs a lot of maintenance, why not grow vegetables on a roof instead? A small example of this can be seen in the park on top of Casa Clementi's car parks. Beautiful environments can be built almost everywhere and in every form, so long as there is a will to do it.

Successful implementation of urban greenery requires practical applications, thorough preparation of designs, plans and submissions, deep levels of coordination between all parties, quality construction, supervision and management. Design and management can generate effective use of spaces, in terms of increasing the amount of space on buildings and giving every space a useful function and maintenance to ensure longevity of the greenery.

One of the most fundamental technicalities for successful implementation of landscapes on top of structures is to incorporate landscape engineering/structural requirements into the building's overall structural design. An example is the South Beach Sky Gardens where the space between the ceiling level of the room below and the floor level above, provided enough soil depth for the roof garden, providing for all the planters and other structures necessary for the rooftop landscape. As such, the space allocated for the rooftop landscape became unusually deep while still catering for the structural needs such as those of the suspended ceiling and services below, the planting areas and the drainage, as well as the occupant load control by raising the soil body. All these landscape structural details were taken into account in the comprehensive dimensions of the building, from the very top to the very bottom. With intelligent design such as this, more beautiful and impactful sky gardens can

Fig. 11.2 Planter encroached by beams; *Credit to* ICN Design International

be created. However, this example is relatively rare. In many cases, the engineering requirements of high-rise buildings rarely cater for the landscape. So, sky gardens are often rather narrow and ineffective spaces with minimal soil bodies, sitting within the footprint of the building, dominated by the lift core.

In the future, as thousands of LRA buildings are built, bad habits have to change. One major issue can be referred to as 'the rise of the beams'. Beams rising up into planters occur when, during construction, it is discovered that the original structural design is inadequate and additional beams are needed but without sufficient headroom to place them below floor level. The beams are conveniently placed inside the planters (Fig. 11.2). The rules of open and full soil- depth planters laid down by landscape architects from earlier design stages are superseded by the sudden appearance of huge beams, a particularly common phenomenon in complex-shaped buildings which require many transfer beams. With the beams cast, and the spaces to accommodate the soil bodies encroached, the entire landscape and planting designs have to be redesigned often causing a major reduction in the numbers of the planned trees. Landscapes have to be reduced and some even totally deleted. This affects not only the LRA and the Green Mark calculations but also the buyers of the property, who did not get the green amenities that were promised to them.

How can we create a more intelligent design for sky gardens to counteract this? What is needed are pre-planning at the initial design stages and commitment to these

Fig. 11.3 Tree planter, big lights and junction boxes; *Credit to* ICN Design International

plans through construction. For example, in the time between design and construction, a designed sub-soil drainage system can pass from a design with adequate drainage, where the specification is for large volume drain penetrations through beams to only two tiny holes that will certainly fail. Cutting extra holes through 3 m -wide beams after casting is not going to happen. Waterlogging and dead plants are the result. Our decades-long campaign to stop electrical contractors mounting junction boxes on rusty bits of metal and sticking them in the most visible places still has not succeeded (Fig. 11.3). Old habits die hard. To get trees on top of structures, a difficult and heavy process, the loading, lifting, programming and cost logistics must be prioritised but rarely are. The procedures to get those soil volumes and heavy, sensitive plants up to the top are rarely taken seriously, resulting in inefficient and dangerous manual lifting.

While new green spaces are important for quality environments, so is the original land and nature removed. In standard construction practice, excavation works remove natural ground permanently. However, careful control of the excavation coupled with the creation of landscapes on top of structures built at the site can produce quality environments. A case in point was the massive excavation of land for the building of the Resort World's basement and car park (Fig. 11.4), which can accommodate nearly 5,000 cars and is 15 m below sea level. This excavation was carefully curtailed such that the adjacent piece of natural forest was kept and together with the new landscape on top of the car park's concrete roof resulted in a beautiful environment of both the old conserved forest and the new multi-level landscape.

Pre-conceived ideas about plants on roofs need to change. With respect to trees, architects, in particular, should not make demands for tree species that can grow to a huge size on top of roofs. For instance, *Bucida buceras* and *Schizolobium parahyba* look aesthetically pleasing when they are small, but both can grow up to 600 mm in trunk diameter and up to about 30 m in height. Such huge plants can block drainage systems with their massive roots and will pose safety concerns if they blow over in the high winds experienced on top of buildings. Additionally, plants need ample soil volumes lest their roots become bound up by being confined. Naturally, trees require ample volumes within soil bodies. A straightforward solution is to plant smaller trees. There are many small trees which are beautiful, have flowers and attract birds and butterflies. The advance-growing of trees makes a lot of difference in ensuring high-quality plant materials at the required size and condition. To provide for future developments such as the Jurong Lake District, trees have to be grown and every single one of them has to be lifted up to the top of buildings together with the soil they need. The Government's proposal to release 100 hectares of land for the development of landscape nurseries will help to supply the hundreds of thousands of trees that will be placed on and around buildings, parks, streets and plazas in Singapore over the next 50 years.

We must not forget green roofs. There are not enough extensive green roofs in Singapore. Green roofs insulate buildings, are cost-efficient and very light. They occupy only 100 mm of soil depth and provide excellent insulation to upper floors. Air-conditioning can be reduced. Not only do green roofs look good from above but they provide a habitat for birds and bees. The use of green roofs by the Housing Development Board for many of their developments has been encouraging.

Important to successful landscapes in the sky and less technical: place-making—the implementation of programmes to activate and connect spaces for users. Too often, decks which are beautifully paved and landscaped are devoid of people because they are not amenable and usable. The Outdoor Refreshment Area and Gross Floor Area regulations often prevent cafes, outdoor dining and the like from being built and operated. To prevent the 60% of LRA for a hard landscape from being useless, programmes for using these outdoor spaces need to be worked out in advance by owners and operators. A case in point is the linkage between Marina Square and The Millennia, which provided a large space with above street-level connectivity but was not effectively activated. As a result, the public space has remained dead for over 30 years since it was built and commercial ventures like cafes and bars have all

Fig. 11.4 Conservation of forest edge, Singapore; *Credit to* ICN Design International

failed there. On the other hand, at the hotel just next door, there are well-maintained roof gardens and thus, the presence of attractive gardens is guaranteed. The vibrant living city does not happen on its own. It has to be built well and with purpose, and then stimulated and managed.

Mr. Henry Steed is a Chartered Landscape Architect (UK) with 45 years of experience, 38 in the Asian tropics. He is Design Director of ICN Design International, an award-winning firm of landscape architects based in Singapore. Henry's years of working in Asia and the Middle East have gained him a reputation as a leading landscape designer. His experience extends through tropical, arid and temperate zones, in urban, rural and natural environments. The making of fine landscapes, to Henry, involves sensitive artistry and hard-nosed technical knowledge of construction and horticulture. These skills are used to transform urban hardness into attractive landscapes for people to enjoy and to stay in touch with nature. Henry has created every kind of landscape, from nature conservation and nature trails, to gardens for hotels and housing, parks, campuses, zoos, streetscape, airports, hospitals and civic design for high rise city centres. Passing on his experience, he teaches NUS students and his staff, encouraging them to develop expertise and create their own visions for the future. Henry has written and illustrated the book titled "Greening the Vertical Garden City" (published by URA and NParks), which deals with the techniques to design and install landscapes at every level in the high rise city.

Chapter 12
From Biophilic Architecture to Biophilic Cities

Tan Shao Yen

'From Biophilic Architecture to Biophilic Cities' is a summary of some of CPG's inquiries into the notion of biophilic design through practice. Many of us are familiar with the term 'biophilia'. It is self-explanatory in that 'bio' refers to living things, and 'philia' means love or affinity. The concept started as the hypothesis that human beings have an innate affinity towards nature and living things. The term was first coined by Erich Fromm in the 1970s and later made popular by Edward O. Wilson. As a multidisciplinary practice delivering built environment solutions, we at CPG, through our course of work as well as personal passions and curiosities, are interested in topics relating to how the built environment and nature can be more interconnected and interdependent for a long time. When CPG was still the Public Works Department (PWD), we were involved in many nature-related projects, for example Chek Jawa Wetlands Park and various projects in the Botanic Gardens. In those days, the term 'biophilia' was still alien to us. But over time, as more and more people from various disciplines within CPG became involved in projects seeking integration with nature, we started an inquiry into biophilic design through our projects and hence through practice. These projects are on different scales, from buildings to neighbourhoods and even to urban districts and are in different geographical locations in Asia, including Singapore and China.

Why do we pursue biophilic design? For us, biophilia is not only a hypothesis. There are compelling reasons and supporting evidence that biophilic design has important benefits. Many scientific studies have validated how it contributes positively to personal well-being, productivity and societal relationships. For example, being in proximity to nature helps emotional restoration, lowers stress levels and improves the possibility of cardiac deceleration. Natural environments facilitate and provide opportunities for physical activities. They foster community cohesion, social interaction and eventually a sense of stewardship towards nature. The resulting increased individual and social well-being also positively impacts productivity, which contributes to socio-economic developments, which leads to yet more biophilic developments, creating and reinforcing a virtuous cycles and positive feedback loops.

T. Schröpfer and S. Menz, *Dense and Green Building Typologies*, SpringerBriefs in Architectural Design and Technology,
https://doi.org/10.1007/978-981-13-0713-3_12

Fig. 12.1 Nanyang Technological University School of Art, Design and Media, Singapore; *Credit to* CPG Consultants

At CPG, biophilic work is part of our integrative design approach. In suitable projects, we seek to enhance them with various degrees of biophilic design by focusing on a few key steps, subject to the stakeholders' buy-in. The first or most basic step is to create or sustain a visual connection with natural features. Many architects do this instinctively as part of their design ethos. The next step is to introduce the physical properties of natural light and ventilation, water features and other natural patterns into the interiors of the building where the users can experience them. In this process, more considerations are required, including the maintenance and safety of the natural elements, the selection of suitable plants and materials and user engagement. The third step is to connect the interior and the exterior physically and as seamlessly as possible to create opportunities for inhabitants to enjoy the exterior easily. At this juncture, the integration of our transdisciplinary knowledge, e.g. about environmental protection, low impact approaches to construction in natural environments, or biodiversity and wildlife, become essential. Manifestations of these steps can be found in many buildings designed by CPG's multidisciplinary teams.

I share a few completed building projects that demonstrate our enquiry process. Completed in 2006, the School of Art, Design and Media (ADM) at Nanyang Technological University (NTU) in Singapore is a popular, iconic example (Fig. 12.1). The building is located in a natural setting, a wooded valley, which in the NTU masterplan provides a 'green lung' for the campus. In order to harmonise the building with its natural setting, it was first conceptualised as a non-building by 'lifting' the lawn and locating the human activities underneath. To respond to the site, two curved turfed buildings were laid out. They embrace a courtyard that serves as a foyer space and the main entrance to the building.

The key feature of the building is its transparency and connectivity—both in the interior spaces and to the external environment. There is a great sense of continuity in the interior, from the entrance to the main foyer to various spaces inside and onto the turfed roofs. Internal glass walls enhance this visual connectivity and flow, allowing users to look beyond the rooms and thus promoting interaction and creative exchange.

The glass facade of the building allows for unobstructed views to the exterior, providing visual connectivity with the surrounding lush landscape. At night, the building glows like a lantern, allowing activities of the school to be observed from the outside. The changing character of the facade creates dynamism and life for the building as day turns into night.

The green roof reduces solar gains and slows down rainwater run-offs during Singapore's frequent heavy rains. The turfs are irrigated using rainwater that is collected in storage tanks. Moisture retention materials installed beneath the soil help to keep the grass consistently damp, which mitigates urban heat island effects, creating a comfortable thermal environment for the users inside and outside.

In the case of the Khoo Teck Puat Hospital (KTPH), completed in 2010, the design is based on the concept of providing 'a hospital in a garden, a garden in a hospital' as a response to the competition design brief which included KTPH's vision of having 'a hospital as a healing garden' (Fig. 12.2). The design was considered unconventional for a hospital as it organised the buildings around a central garden that opens up to the adjacent Yishun Park and Pond. This is in contrast to more conventionally planned hospitals that give priority to efficient workflows, traffic and mechanical services. Obviously, the hospital workflow and services are mission-critical and needed to must be addresses. However, in response to the client's visionary brief, the design managed to balance biophilia and connection to nature with other important considerations. In this project, the vision of KTPH's CEO Mr. Liak Teng Lit and his team was instrumental in pushing our design approach and resulted in an award-winning project.

The design of KTPH is effective in enabling its public space to function as a community space by extending the adjacent Yishun Park into the project. Park users are drawn into the hospital compound in an almost seamless way and vice versa, patients of the hospital get to view and enjoy the adjacent public park. As part of Yishun Park's natural ecosystem, about 80 species of butterflies have been sighted in the KTPH gardens, attracting both nature enthusiasts and the general public. The roof spaces of KTPH are used for urban farming as well and engage volunteers from the neighbourhood. Interestingly, the Yishun area used to have farms and some retired farmers from the community have volunteered to cultivate the roof farm, generating a sense of community stewardship.

The ideas and vast areas set aside for gardens, landscapes and urban farming did not happen by chance. The Hospital Planning Committee, management and staff were from the former Alexandra Hospital, an early colonial building that occupied a site about three times the size of the new KTPH. The Alexandra Hospital staff community had already built an active culture to cultivate its land with greenery for farming and biodiversity to thrive and well known for its butterfly gardens. This ethos was conceptualised and provided the basis for the design of KTPH.

Fig. 12.2 Khoo Teck Puat Hospital, Singapore; *Credit to* Ang Siew Tiang

In the Ng Teng Fong General Hospital (NTFGH), a project completed in 2015 and set in a more urban context, we took a different approach (Fig. 12.3). The key concept of the project is that every bed should have a view. This approach is based on findings provided by scientific studies that demonstrate that visual connections with the external environment have physiological and psychological benefits and support patient well-being and healing. By adopting a fan-shaped ward layout, each bed of the public hospital is situated next to an external wall with its own window. This allows for personalised control of thermal comfort, air and light. It also allows the patients to visually connect to the external environment. The abundant daylight, cross-ventilation and visual connectivity make it a better work environment for healthcare workers, too. With its fan-shaped ward design and its interesting facade that integrates vertical greenery, many people from the public thought it was a condominium when it first opened. This was desirable, because as part of the urban fabric of Jurong East Gateway, the hospital must play a role in enriching the urban environment.

One of the ongoing inquiries that CPG has is about the Green Plot Ratio. Can high-density developments be approached as an opportunity for more greenery? The Solaris project, in which CPG collaborated with T. R. Hamzah & Yeang, provided very good insights. Completed in 2012 and located in One-North, the building is a 17-storey high multi-tenanted facility with organisations working in the fields of information and communications as well as media science and research. It was stipulated in the tender brief for this project that a minimum number of pocketed green areas were required. However, the client, Soilbuild, was both the developer and

Fig. 12.3 Ng Teng Fong General Hospital, Singapore; *Credit to* Rory Daniel

builder of the project and aspired to do more than that. It supported the architects' vision of a 1.5 km-long continuous greenery corridor that winds up the external envelope of the building, connecting the ground level to the roof gardens (Fig. 12.4). Through this vertical greenery strategy, the Green Plot Ratio achieved was eight times what it needed to be. Clearly, with the right policy, motivation and business model approach, built environment projects that support human use need not always deplete the natural environment; they can simultaneously be conceived as opportunities to intensify nature, albeit through an altered form.

The Gardens by the Bay, completed in 2012, is a well-known biophilic project (Fig. 12.5). Our key collaborators were Grant Associates for the Gardens and Wilkinson Eyre Architects for the two domes. Given that the project is first and foremost a park, it goes without saying that it is biophilic in nature. It is built on reclaimed land that has been transformed into a public park for our local communities, a destination for visitors, and a natural environment to support our ecosystems. When we look closer, the intriguing aspect about the park design is that it is a system of systems. Multiple inter-related sub-systems are working together, including the energy and the water system. The project has natural ecological cycles and material flows and as a result of all of these features a human activity system. And, like in an ecosystems, these sub-systems support each other in an integrated manner. For example,

Fig. 12.4 Solaris, Singapore; *Credit to* Albert Lim

biowaste such as fallen leaves and branches that are collected from Singapore must be incinerated and gotten rid off every day. The incineration is carried out at the Gardens' energy centre, which converts the heat generated through a desiccant system to provide cooling for the two domes. To put it simply, biowaste from the greenery of Singapore is not regarded as waste but as a useful resource for the Gardens. Similarly, water from Marina Bay is channelled into the park as a resource for irrigation and at the same time the water plants of the Gardens are designed such that they ensure that the water quality at the point of discharge back into Marina Bay is of a quality than at the point of intake. The design of these systems requires knowledge and inputs from

Fig. 12.5 Gardens by the Bay, Singapore; *Credit to* Harshan Thomson

many different disciplines. The ability to manage the integration of their knowledge is equally crucial. Again, this is a project driven by a client organisation with a strong vision and with knowledge of biodiversity that allows for it to become a success. In this case, we are attributing important ideas to Dr Tan Wee Kiat and Mr Kenneth Er and their teams from NParks.

Beyond building design to there is environmental design. In Singapore, various programmes have been created by Government agencies including the URA, PUB and NParks for the integration of buildings into the green and blue networks of the City. One of these programmes is PUB's Active Beautiful Clean Waters (ABC Waters) programme. It is an integrated water resource management scheme to treat water holistically. In 2010, CPG was part of a consortium that completed a master plan for the PUB with extensive involvement from other Government agencies. This master plan covered approximately one-third of Singapore in the northeast and part of the scope included the identification of ABC Waters projects as pilots for implementation within the area.

One project example is the Lorong Halus Wetlands (Fig. 12.6). Completed in 2010 and located next to the Serangoon Reservoir, the manmade wetlands were designed to treat leachate of an old abandoned landfill situated next to it. When it was decided that Sungei Serangoon would be converted into a reservoir, the discharge from the landfill had to be dealt with. To address the problem, a wall was built along the side of the reservoir, preventing direct landfill discharge into the river. However, as rainwater falls unto the landfill, the water level rises and there was a risk of water flowing into the reservoir. The solution was to capture the overflow and treat the water quality to

Fig. 12.6 Lorong Halus Wetlands, Singapore; *Credit to* CPG Consultants Pte Ltd

a safe level before it is discharged into the reservoir. Instead of using a traditional treatment facility, the ABC Waters approach was not only innovative but also brings more benefits for the local community. The Lorong Halus Wetlands project filters the water through a series of filtration steps, using bacteria and plants which ensure that by the time it is discharged into the reservoir, the desired quality is achieved. At the same time, the natural plants were selected to create a wetland habitat, attracting migratory birds and people to use the project as a public space. This is yet another example that emphasises the growing need for multidisciplinary design approaches, in this case one that integrated landscape design and environmental engineering and provided water quality improvement technologies and hydraulic calculations.

Another project example within this ABC Waters master plan is the Sengkang Floating Island (Fig. 12.7), also completed in 2010. It is located in Sungei Punggol next to the Anchorvale Community Centre. There is a sports and community centre that is located on one side of the river and the opposite side has a fruit park. To integrate the use of these two spaces for the community, there was a need to connect them. One of the project requirements was to minimise the intervention regarding the river, particularly its capacity. An innovative approach was taken by designing bridges that are supported by floating platforms that form a floating island. The bridge and the floating platforms were fastened onto pylons anchored into the river bed in a way that permits vertical movement to accommodate tidal changes. The floating island was designed to support human live loads, play equipment as well as water plants with bioremedial roots that are able to remove excess nutrients and improve water quality.

Fig. 12.7 Sengkang Floating Island, Singapore; *Credit to* CPG Consultants Pte Ltd

There is a limit to what each individual development can achieve. This is the reason why many systemic benefits will have to be achieved on the larger urban scale, for example through a district cooling system or an urban transportation system. Hence, more recently, our inquiry into biophilic design is taking into account our master planning and urban design projects.

One of the projects that we have undertaken in Singapore is the Kallang Riverside master plan. The project was led by the URA and multiple other Government agencies participated in it. One of the project's unique features is that the ABC Waters programmatic considerations were incorporated. The design strategy was considered holistically, from individual parcels to the district level. The benefits were explored quantitatively in terms of cost-saving and socio-economic factors.

Another project we participated in was a master plan design competition for Danang City, the third largest city in Vietnam. Danang has a very important natural heritage, the Han River that runs through the city. Because of its economic success, Danang has undergone rapid urbanisation and is quickly becoming a concrete jungle. Our design proposed a blueprint for the creation of biophilic environments through various strategies.

In summary, in our biophilic vision for a better world, projects are designed and built not only to fulfil the human, economic and social needs but also to enrich the natural environment. On the district level, blue and green infrastructure are put in place so that the environment, the connection between buildings and the in-between spaces are biophilic. Our projects, each with its own characteristics are connected to form the city or even a mega-city region through combining elements such as rivers and green corridors whereever possible. In many ways, this vision is inspired by where we are, in Singapore, a city that is continuously experimenting with interesting policies and incentives to encourage urban greenery. Notwithstanding the policies and incentives, the visions provided by developers and clients are paramount. Without visions and objective-setting, we cannot expect any success. Finally, biophilic environments start with biophilic design. Hence, for designers, it is important to develop abilities to approach design in a biophilic way by overcoming knowledge silos, by learning how to integrate knowhow, by engaging stakeholders and by developing work processes that support collaboration.

Mr. Tan Shao Yen As the Chief Executive Officer of CPG Consultants, he is responsible for the business and operations of CPG Consultants and its overseas offices in China, India, Vietnam, Macau, Philippines and the Middle East. He is supported by a multi-disciplinary community of architects, engineers, planners, designers, and specialists. Together, they believe in following an integrative approach to providing holistic solutions, by collaborating through integrated teaming involving clients, designers, consultants and other stakeholders; and applying research, cross-disciplinary knowledge and technology in the design process. Trained as an architect, Shao Yen has over 25 years of experience designing and delivering projects from a broad range of building types in the public and private sectors and in different climatic and cultural regions, including Singapore, China, Malaysia, Vietnam, and the Philippines. Some of his ideas were selected, exhibited and published as part of the Singapore Urban Redevelopment Authority's 2010 '20 under 45: The Next Generation'. In 2014, he was recognized as the 'Green Architect of the Year' by the Building and Construction Authority Singapore Green Building Council. Mr Tan Shao Yen currently also serves as the President of the Singapore Board of Architects.

Chapter 13
Emulating and Replicating Ecosystems Ability to Provide 'Ecosystem Services' by the Built Environment

Ken Yeang

The existing built environment is ecologically dysfunctional in its relationship with the natural environment and this needs to be addressed. This is the most compelling issue confronting architects, planners and engineers today—how can we reinvent, redesign and remake our built environment to make it resilient, durable and sustainable? The contention here is that we must learn from nature as our mentor and as our model. We must work and design with nature and not against it.

What is 'designing with nature' that is referred to by some as green or sustainable design? There are many definitions and approaches, but the one used here is designing for an effective biointegration of our built environment with the natural environment. The contention is, if we are able to biointegrate everything we as a human society make and do in the natural environment in a seamless and benign way, then there will be no environmental issues whatsoever. This is, of course, easier said than done but this is the challenge of ecological design.

We will find similarity in what designers do with what surgeons do in the prosthesis. In prosthesis, a device like a prosthetic artificial arm or other implants are attached to a host organism. For this connection to work, it depends on effective biointegration. If the artificial device does not integrate seamlessly, benignly and effectively, with the host organism, then either the host organism's health is affected or the prosthetic device breaks down.

By analogy, the built environment is like a prosthetic device. It is artificial, synthetic and human-made. The crucial question then is what is its equivalent host organism that it must seamlessly and benignly integrate with? The answer is the biosphere. If we are able to seamlessly and benignly integrate everything that human society makes and does in the biosphere in the same way that we biointegrate a prosthetic device with its host organism, then there will not be any environmental issues. This is what ecological design or green design must achieve.

What is effective biointegration? Effective biointegration requires the built environment to be ecosystem-like, to become a human-made ecosystem as an extension of nature, being an extension of nature and not being a separate or disparate entity. Like an effective prosthetic device, the built environment must become a hybrid

T. Schröpfer and S. Menz, *Dense and Green Building Typologies*, SpringerBriefs in Architectural Design and Technology,
https://doi.org/10.1007/978-981-13-0713-3_13

ecosystem. In this way, we avoid a dislocation between the natural environment and the built system.

There is still much to be done in prosthetic biointegration as the relationship between the prosthetic device and its host body is still not perfect. For instance, with a prosthetic arm, the junction between the prosthetic device and the host body's nervous system, muscular system and other circulation systems remain similar to an open wound such that the host organism as an individual human may need to be constantly on antiseptics.

Ecologists see the earth as covered by a thin film called the biosphere where organisms live. Within the biosphere are units called ecosystems, which consist of communities of plants and animals with their physical environments, with biotic constituents and abiotic constituents that act together to form a whole system.

Since the Industrial Revolution 200 years ago, human beings have become the single largest and most powerful species on the planet. Humans have radically changed landscapes, used huge amounts of non-renewable energy extracted from the planet, and now we are even able to change the global climate to the detriment of the planet.

The human community also makes things and in effect, it makes many more artefacts than any other species in nature. Society's built environment structures and artefacts are all synthetic and potential waste. Their production requires the generation of energy from non-renewable sources and emits waste. The things that humanity throws away produce emissions and waste in the ecosystems. These are the crucial causes of environmental disruption and damage that need to be rectified. Nature has survived over 3.8 million years but the current impairment to the planet is so extensive that it has been predicted that this millennium may be humanity's last. This is the rationale for the urgency in our remaking of humanity's built environment and restoring of the planet's natural systems.

We need to ask, what if our cities as our built environment become human-made ecosystems? If this can be effectively achieved, the natural and the built environment become fused and blended into a single system and not separate conflicting ones.

How do we make our built environment into constructed ecosystems? Achieving this requires our built environment to emulate and replicate ecosystem attributes in a process called 'ecomimesis' (© Ken Yeang) as the biomimicry of ecosystems. In other words, if ecosystem properties and attributes can be emulated and replicated, then our human-built environment becomes a hybrid constructed ecosystem as an 'eco-cyborg' to biointegrate, to become an extension and a part of nature.

What are the ecosystem attributes that we have to emulate in the built environment? The most important is the ability of ecosystems to provide ecosystem services, whereas other attributes include the biological structure of ecosystems, the biointegration within ecosystems, ecosystems homeostasis, among many others. To understand and describe each of these many attributes requires outlining a larger body of work, but a few examples can be given.

The biological structure of ecosystems consists of the biotic constituents including plants, animals, microbes and all organic life; and the abiotic environment including the inorganic physical environment, climate and geology. However, today almost everything made by humans is abiotic, with less biotic components in the built environment. If the built environment is to imitate and replicate ecosystems, then it must have a complement of biotic components that need to be biointegrated with the built environment.

This is not biophilia, which is based on human perception and well-being. Biophilia architecture contributes essentially no biological value to the built environment except as a side benefit. The reasons for putting vegetation or organic life into buildings and built environments is not just for decorative or perceptual reasons, nor climatological such as for lowering the heat island effect on cities, but also ecological to make the city much more ecosystem-like by having ecosystem attributes, including the enhancement of its biodiversity and other functions.

Biotic components can be put into buildings in a variety of patterns—(i) by juxtaposition, so that all vegetation is placed in one location, (ii) by intermixing as dispersed patches within the built form, (iii) by enhanced intermixing so that the vegetation is in a 'stepping-stone' pattern and (iv) by an intertwining pattern, extended if possible to its bioregion and to its hinterland in an ecological nexus. Ecologists contend that a continuous pattern is the best as it enables species migration and interaction, but more importantly, by being linked, it enables a larger pool of shared natural resources within the ecosystem that encourages greater stability and overall supports a higher level of biodiversity.

These patterns can also be applied on a master plan scale. The juxtaposition of having one large patch of greenery, such as Central Park in New York City is less effective in providing ecosystem services to the entire urban realm, with regard to the significant intensity of the inorganic mass of the buildings in that city. An example of intermixing translated to a master plan can be found in Central London, which has a series of green squares such as Bradford Square, Tavistock Square and Russell Square. Stepping stones pattern of greening can be created in the landscape to allow some connectivity. Finally, the continuous greenery on a building in the form of ecological corridors better serves as a series of connected habitats and as wildlife corridors with eco-fingers.

Which of these patterns is preferred? The continuous linked vegetation on buildings is better and should be adopted, and when there is the usual clearing and fragmentation of vegetated and forested land for urban development, this fragmentation needs to be repaired to make the ecosystems whole, where appropriate. This is one of the missions of ecological architecture.

The first step when designing a building is to understand the level of biodiversity around it. Singapore is a city at about 1-degree latitude north of the equator, in an area where naturally we would find tropical rainforests. This means the locality has some of the highest levels of biodiversity in the world. The intention in our design for the Solaris building at One-North was to bring the vegetation as a landscaped ramp from the lowest levels, at the sides of the building, in an ecological nexus up to the roof gardens (Figs. 13.1 and 13.2).

Fig. 13.1 Solaris, Singapore; *Credit to* Albert Lim KS

This design idea appeared earlier in a scheme for the EDITT tower at Waterloo Road in Singapore in 1999 from which the idea of a landscape ramp for vegetation was developed. This concept was finally implemented in 2005 in Solaris with a ramped vegetated walkway on the facade as a 'linear park' where visitors and users

Fig. 13.2 Solaris roof gardens, Singapore; *Credit to* Albert Lim KS

can go from the ground floor all the way up to the top of the building. It is over 1.3 km long, probably the longest linear park in a building in the world.

At the corners of the building, this linear park unfolds into mini-plazas so people outside can interact with those inside the building. The idea of a sky court, as a plaza that punctuates the building is a recurring theme in many of our projects, such as in the sky-courts in the Menara Mesiniaga, which received the Aga Khan Award. In some of these buildings, there are also 'eco-cells' as cellular devices that bring rainwater, natural ventilation, daylight and vegetation to the insides. The requirement for Singapore's Green Mark is to achieve a making of six points; however, Solaris was awarded 12, which is twice what was standard at that time. The building also has a naturally ventilated central atrium with a bridge linking the two towers within the built form and a diagonal light shaft that provides views from the bottom to the top of the building.

Our work does more than just putting landscaping on buildings and, not as in the case of some architects, using planting as simply decoration. Our projects create habitats within the built form to enhance biodiversity, which is ecologically beneficial to the locality beyond just planting. An example of this is approach is the Suasana Putrajaya building in Malaysia, completed recently, located near the bridge and on the axis of the Millennium Monument. The habitats in this building were matched with selected fauna species (non-hazardous to humans) and native to the area, to bring them back to the existing site that is bereft of wildlife. A biodiversity matrix was created to become a guide for designing the habitats for the landscape architecture of the building. An example of a habitat is between the two blocks.

Fig. 13.3 La Reunion masterplan design; *Credit to* T.R. Hamzah and Yeang

The design of the east side and west side of each block responds differently to the sun path. The building is also designed to look upmarket with a 'diamond' aesthetic achieved by its faceted shape. The fritted-glass outer-skin sun shading has a Malay 'songket' pattern for its vital cultural link.

Generally stated, the inclusion of biotic constituents in the urban environment further enhances its ability to provide 'ecosystem services' but this green eco-infrastructure needs to be woven into the city's fabric. Provision of ecosystem services is one of the key ecosystem attributes to be emulated by our built environment. Nature provides ecosystem services for free without human intervention.

This is demonstrated in our master plan for a site on the island of La Reunion, east of Madagascar (Fig. 13.3). The vegetation is laid out between the urban strips and grows upwards towards the hills, interwoven with the urban fabric. The master plan shows how we could biointegrate vegetation and nature in the city, not as isolated patches, but as a series of green eco-infrastructures interconnected to the hinterland as a series of linking eco-fingers.

Can we remake our built environment the way nature does it? Yes, but it requires effective biointegration in the remaking of the city and built environment as 'constructed ecosystems'.

Dr. Ken Yeang is an architect, planner and ecologist who is best known for his signature green architecture and masterplanning, differentiated from other green architects by their authentic ecology-based approach, by their distinctive green aesthetic and green performance beyond conventional rating systems. He trained at the AA (Architectural Association, UK) and received his doctorate from Cambridge University (UK) on ecological design and planning. His key buildings include Solaris (Singapore), Menara Mesiniaga (Malaysia), Spire Edge Tower (India), Genome Research Building (Hong Kong), Great Ormond Street Children's Hospital Extension (UK). He is Principal of T. R. Hamzah & Yeang with offices in Malaysia, UK and of North Hamzah Architectural and Engineering Company in China. He is a recipient of the Malaysian Institute of Architects Gold Medal, the Government of Malaysia Merdeka Award and the Architectural Society of China Liang Sicheng Award 2016. The UK Guardian newspaper named him as one of 50 individuals who could save the planet, and he is named by CNN as the leading architect in ecological design.

Chapter 14
Greenery in Commercial Buildings: Enhancing Returns for Investors

Megan Walters

Commercial office buildings in a city normally make up approximately 25% of the built stock. In Singapore's Central Business District (CBD) this percentage is higher, about 70–80% of the built stock. Without consideration for the surrounding environment in building design, these commercial offices can have a negative impact on visitors and residents, which in economics is known as a negative externality. The question that investors are increasingly asking themselves is, how can they prevent these negative externalities and generate a clean income stream from those commercial buildings? What role can greenery on a building play to not only mitigate these externalities, but also to enhance income streams and provide positive externalities?

To better understand what occupiers want from buildings, it is important to know the concept of the urban economic cycle and the role it plays in shaping cities. The first urban economic cycle was the advent of steam and manufacturing which changed the shape of cities. The second cycle was electricity, this made cities much more vertical because power could be generated for lifts and because electric lighting made them safer. The third cycle was about cars and how this created the rise of the suburbs. Today the fourth urban cycle is about the future of work, and it has five main components that can influence how a city is shaped: human experience, digital drive, continuous innovation, operational excellence and financial performance. Greenery on and around buildings can significantly influence human experience, and therefore influence the new urban economic cycle.

In the rental market, large-scale corporate occupiers will rarely pay a substantial amount over the base market rent value for office space. Therefore, the competition between rental buildings for occupancy is driven by the experience that a building can give to occupiers' employees. A better experience for occupiers' employees for the same rental price as a less experiential building can minimise building vacancy rates and tenant turnover, enhance income streams, which in turn provide a higher operating income, leading investors to pay more per square foot for a building. This has made the word 'experience' very important in commercial real estate; and there is a good understanding from investors on how plants in a commercial building can contribute to occupier experience.

T. Schröpfer and S. Menz, *Dense and Green Building Typologies*, SpringerBriefs in Architectural Design and Technology,
https://doi.org/10.1007/978-981-13-0713-3_14

The positive effect from occupier experience, increased tenant occupation and retention rates also reduces investment risk: A green building being close to full occupancy with stable tenants makes an event where rental income is not being made from empty offices less likely. Importantly, there is also an understanding from occupiers about how greenery can help create a work environment in which you can attract and retain a large pool of talented employees over time, based on it providing a positive experience. Once you start to get a pool of talent together, there is agglomeration. The talent will attract other talent to that place. This will then further encourage other occupiers (for example tech companies) to move to that area to tap into that talent.

An example of investors considering greenery as a way of enhancing human experience is the Embassy Business Park in Bengaluru, Karnataka, which is co-ventured with Blackstone, one of the biggest global investors. Throughout the development, greenery has been incorporated as places to create and grow vegetables and provide locations to enhance the environment for employee experience. Similarly, elsewhere around the Asia-Pacific region other investors believe that greenery can enhance human experience. For example, in Bangkok a new development set for completion in 2021 has a total land area of 167,000 sq m which will have 80,000 sq m of greenery or open space. This is also an example of how a Singapore property investor (Frasers Property Holding) has influenced buildings to be constructed with extensive greenery in the wider region. As a developed city in the Asia-Pacific region with a history of urban greening, there is much Singapore can teach the rest of Asia about the value of building and urban greenery.

Another important source of future investment in green buildings is from investors that are required to meet environmental, social and governance (ESG) requirements in their investments, for public perception and due diligence purposes. In this market, green buildings compete with bonds. However, green buildings are less likely to have clearly defined ESG benefits and can therefore be less attractive to investors than certified ESG bonds. An important step forward to address this would be formalising the ESG benefits of green buildings. This would convince investors that they can mitigate negative externalities and generate a clean income stream.

Over the past few years, since our global financial crisis, there have been significant increases in the volume of investment dollars going into real estate, particularly into commercial real estate. Large pension funds, which historically have invested approximately 5% of their portfolio in real estate, are now investing 10–15% into commercial real estate. If this weight of money going into buildings is used to enhance the wider environment, then there must be an understanding of why investors are increasing the volume of money they put into real estate. This is where real estate services companies such as JLL that employs 78,000 people globally and has 57 billion dollars worth of assets under management, primarily in commercial real estate, can play a role.

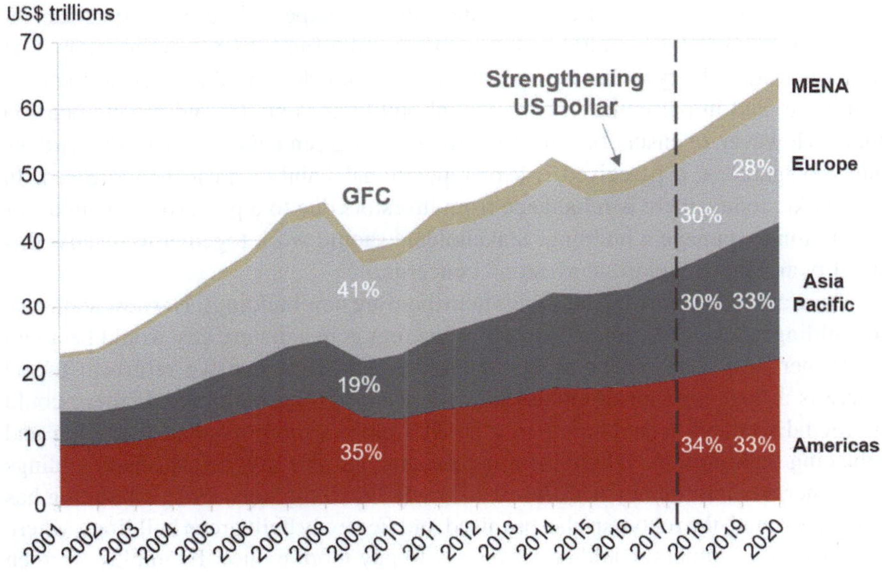

Fig. 14.1 Growth in real estate market per region; *Credit to* JLL Asia-Pacific

The reason why greenery on buildings could potentially be important in the Asia-Pacific region is the size of the real estate markets and the potential for future growth. Even though the whole of the Asia-Pacific region has around half of the world's population it only accounts for about 30% of the global real estate value, with Europe and the United States accounting for around 30% and 34% respectively. The growth in Asia-Pacific GDP output will cause big international investors to shift and allocate more capital in the region as evidence shows commercial real estate in the Asia-Pacific region is expected to increase by 3% in the next 3 years to 33% in 2020 (Fig. 14.1). From any perspective, more money is going to flow into Asia-Pacific real estate, both from home grown wealth or investment in other countries, whether these are Korean pensions funds, Malaysian pension funds or Chinese state-owned companies wanting to invest. Even today, there is expected investment from Japanese Government pension funds into commercial real estate in places outside of Japan; and there is precedence for investment in green buildings within the nation, which could lead them to invest in these buildings abroad.

Investment in green real estate in the future is expected across a spectrum of formal state-owned companies and sovereign wealth funds to private investors. Out of all investors, the type that is expected to invest much more than they do today in real estate and in particular in commercial buildings is public and private pension funds. However, to ensure that any investment is in green buildings with sky gardens and roof gardens, a perceived risk of reputational liability must be addressed. In the market today, there is reluctance from investors due to a perceived potential for vegetation to damage a building. Stakeholders should work together to discuss this liability and how to address investors' concerns.

There are many more old buildings than there are new buildings. Net new additions to building stock across many markets in the region in a mature city would be about 3–5% per year. Can old buildings become green? How can we refurbish them? There is a huge amount of embedded carbon in existing buildings. If these could be revitalised, this would result in a smaller carbon footprint than removing and replacing buildings and overall be more sustainable. The way refurbished buildings can generate more money is different to newly built buildings. Once a building has been renovated, the rents can also be raised, unlike newly built green buildings, where commercial occupiers would be less likely to pay more money. Examples of green refurbishment exist in Spain and Switzerland amongst other countries. Essentially the current reason for increased rent demand is improving aesthetics. If a building has large areas of plain concrete wall, then adding plants and shrubs can improve the way they look, and they can become more visually attractive. In the Asia-Pacific region, 130 billion dollars worth of commercial real estate is bought and sold every year. If someone is selling property, they may want to refurbish it. Therefore, talking to property owners who are in this position about the benefits of greenery on buildings is important as they may consider retrofitting greenery.

Incentives are a hugely important way to change the status quo and encourage new things to happen, which has been the case in Singapore. The National Parks Board Skyrise Greenery incentive scheme funds up to 50% of installation costs of rooftop and vertical greenery; and the Urban Redevelopment Authority offers bonus floor area for conversion of up to 10% of a buildings floor area to sky or roof gardens. These kinds of incentives can play an important role in increasing investment in green buildings and facilitate great cities moving forward.

Dr. Megan Walters is international director and Head of Research at JLL, leading a team of 170 researchers in Asia Pacific, part of a team of 300 researchers globally. The team provides research thought leadership, data collection and analysis across the markets, including JLL's Real Estate Intelligence Service, considered the industry's leading market research service for real estate. Dr Walters is a Fellow of RICS with a PhD from the University of Hong Kong in Institutional Economics of Real Estate. Megan has an interest in the issues of multiple ownership property (strata title) and property rights on urban environments and city development. Megan joined JLL in 2010 as Head of Research, Asia Pacific Capital Markets, a role responsible for advising investor clients and generating publications on office, retail and industrial real estate trends in capital markets. She is also a member of the JLL Global Research Executive with responsibility for the Global Office Leasing Group. Megan sits on the research committees of ANREV and APREA and is the ULI

Women's Leadership representative for Asia Pacific on the WLI Global Committee. Megan held an Adjunct Associate Professor position at National University of Singapore and served on the RICS International Governing Council from 2009 to 2015.

Chapter 15
Green Buildings and the Singaporean Homebuyers

Nicholas Mak

Ten years ago, real estate developers and agents rarely considered using a green message to sell condominiums or other developments. Instead, some of the selling points used to market real estate were the location of the property, amenities and facilities within the development, and the lifestyle that it can provide.

In recent years, developers and consultants have been increasingly promoting green features as a part of the condominium lifestyle, amenity and facility choice, merging these two concepts together. Images showing individuals interacting and using facilities such as swimming pools and gyms among plants and animals, are now often included in real estate marketing collateral. Green walls, plants and eco-friendly symbols are abounding in modern property showrooms.

This raises the questions, what do homebuyers see as 'green' features and are they a priority when selecting new properties to buy; do homebuyers believe the green message real estate developers are trying to sell; and are homebuyers willing to pay more for green real estate? To find answers to these questions, we conducted survey with over 100 respondents, all of which were living in Singapore and most were within the family formation age (Fig. 15.1).

Respondents in the survey saw home green features as energy-saving equipment and appliances, water-saving equipment, the orientation of the dwelling units (is it exposed to the heat from the sun) and insulation to prevent loss of cold air. The most important of these were water-saving features and the least important was housing orientation (possibly due to a misunderstanding of its benefits). Overall, respondents mentioned that they would be more willing to pay for home green features, if they see financial saving, such as savings in utility bills.

Outside the home, respondents saw green features as solar panels, rainwater collection systems, waste management systems, landscape greenery and growing food and herbs (urban farming). The most important of these were greenery enhancements: respondents believed that enhanced greenery could provide health benefits, and also be good for the environment. The least important was growing food and herbs (urban farming). Not surprisingly, a large percentage of respondents felt price, location and size of a home were the most important reasons why they would buy a condo-

T. Schröpfer and S. Menz, *Dense and Green Building Typologies*, SpringerBriefs in Architectural Design and Technology,
https://doi.org/10.1007/978-981-13-0713-3_15

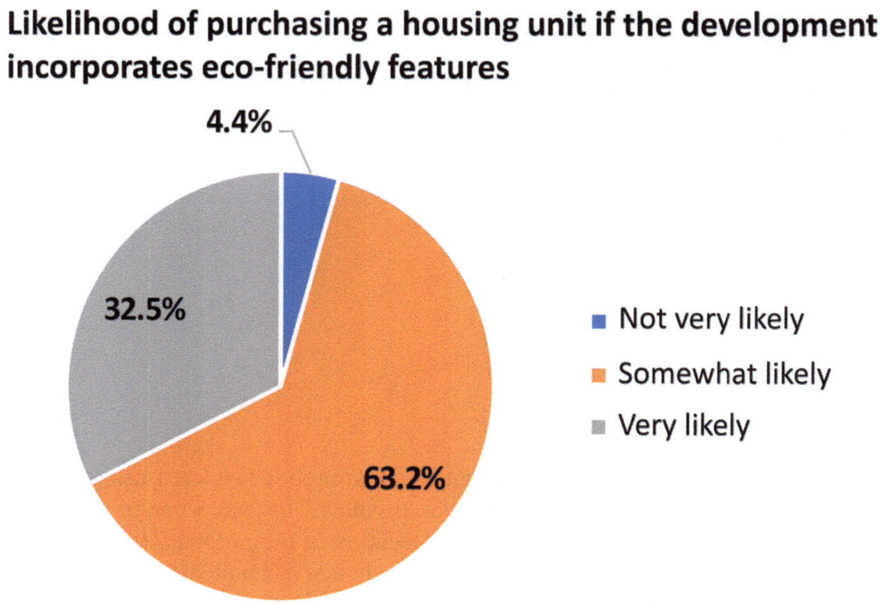

Fig. 15.1 Green features and purchase likelihood; *Credit to* SLP International Property Consultants

minium. Of lower consideration was the availability of green features, although 20% of respondents said it was very important and 50% said it was somewhat important.

How much do individuals buy into the green condominium message? Overall respondents had doubts over developers' and consultants' green marketing claims: only about 8% of them stated they were very convinced, while about 72% said that they were somewhat convinced and about 20% said they were sceptical. By age group, individuals 51–60 years old were less likely to be convinced of green marketing materials. This is significant because those in this age group are more likely to earn more, upgrade their property and be in the market for a new development.

The same survey also showed that over half of the respondents were not willing to pay any additional money for green features. This was largely because they felt that the costs may outweigh the benefits, especially in the short term. They also felt that greenery costs should be borne by developers, even though eventually this cost would be passed on to the homebuyers.

How did these preferences breakdown in the different age groups? Those aged 41–50 years old were more resistant to pay than those older or younger. By income, again, the group that is unwilling to pay more for green features are those more capable, older individuals earning over $8,000 per month.

Respondents who were willing to pay for green features were asked how much they would be willing to pay with the assumption that green features can save them over $100 in utility bills annually (Fig. 15.2). About 20% would pay nothing; about

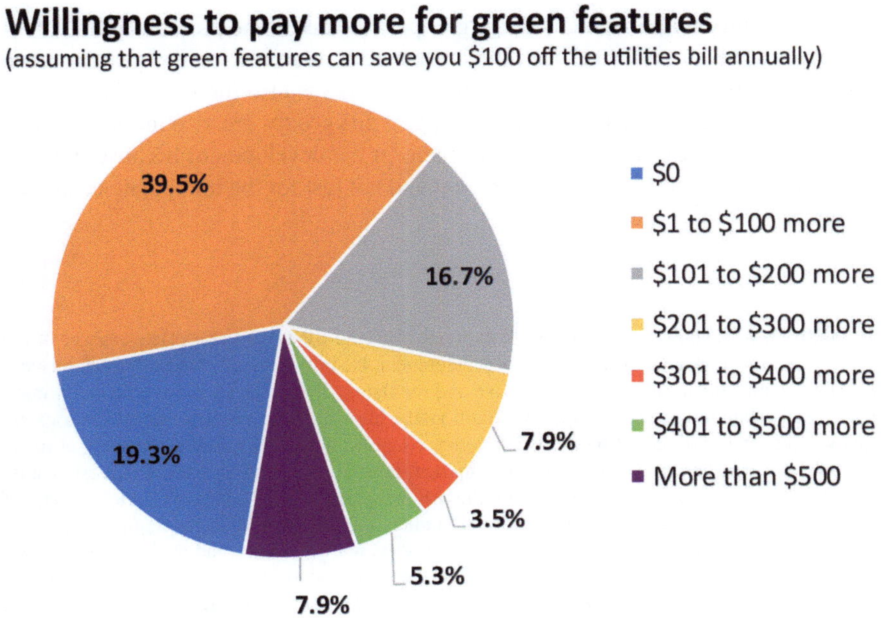

Fig. 15.2 Willingness to pay assuming $100 bill reduction; *Credit to* SLP International Property Consultants

40% were willing to pay between $1–100; about 17% were willing to pay $101–200; about 20% were willing to pay $201–500; and about 8% would pay over $500.

A potential reason why the majority of the respondents were only willing to pay less than $200 for green features was the overall perception that green buildings did not have a positive financial impact on the individual. Contrary to the willingness-to-pay results, all age groups did show a preference for condominiums with eco-friendly features: about 30% were very likely to do so, about 65% were somewhat likely and about 5% were not very likely to do so.

Some homebuyers had interesting perspectives on the role the Government or developers should play in green building development. One respondent said that the Government should levy some type of tax or higher cost onto non-eco-friendly homes. The justification for this was that non-eco-friendly homes are causing negative externalities to the community and that eco-friendly homes should be subsidized. One other respondent stated that there should be a minimum industry or Government standard for a home to be eco-friendly. Another respondent stated that he would prefer developers to provide green features within a new housing unit as the features were expensive to buy and difficult to install.

Overall, the green message is becoming more and more prominent in the Singapore real estate market. Eco-friendly features are more popular with respondents when they see the immediate economic return in terms of energy and water savings. Greenery enhancements outside the home but within developments are also popular.

The respondents in all income and age groups indicated some preference for a condominium to be green. Older homebuyers are a bit more resistant to pay for green features and more sceptical of their benefits, but this is also the group that is more capable of buying private homes. The youngest age groups are also more willing to pay for green features or green enhancement in the development because they are something that is good for the environment and not just for the immediate economic benefit.

Mr. Nicholas Mak has 20 years of work experience in the Singapore real estate market in the areas of research, consultancy and business development. He provides real estate advisory services to major organizations such as the legislative and regulatory bodies in Singapore, financial institutions, real estate funds and developers. Additionally, his views and extensive commentaries are often sought-after and quoted by the local media. Nicholas has also spoken extensively at many seminars on various aspects of the real estate market. Presently, he leads a team of real estate analysts and consultants to conduct research on the real estate market and advisory services. Some of the other studies undertaken include feasibility studies of land and real estate acquisition, development planning and consultancy, product mix advisory, market studies and analysis. A graduate from the University of Technology, Sydney with a degree majoring in Economics and Finance, Nicholas had also pursued and attained the Master of Science in Real Estate from the National University of Singapore.

Chapter 16
The Role of Ecosystem Services in Making Cities Sustainable

Peter Edwards

The term 'ecosystem services' is used to denote the benefits that people obtain from ecosystems, such as clean air and water, or a cooler environment. In this chapter, I will consider some of the environmental problems that cities face, and how ecosystem services can contribute to mitigating them. I will focus especially on cities of the humid tropics, notably Singapore, and the potential of ecosystem services to prevent flooding and to keep the city cool.

Two hundred years ago, Singapore was a tropical rainforest. At that time, the vegetation regulated many aspects of the environment. When it rained, for example the forest behaved like a gigantic sponge, with every leaf holding a few drops of water; as a result, about 30% of the rainfall never reached the ground. The trees also took up water through their roots, which then evaporated from the leaves—a process known as transpiration. Indeed, to judge from remaining forests in the region, somewhere between 58 and 87% of the rain falling upon Singapore evaporated, either from the leaf surfaces or by transpiration, which meant that flooding was a rare occurrence. Today, Singapore—like other large cities—has much less vegetation and about two-thirds of the rain falling on the city runs rapidly off the hard surfaces. Not surprisingly, flash floods have become a problem, despite the construction of massive storm drains.

Rainforests also regulate the air temperature—in marked contrast to cities, which tend to get hot. This tendency for cities to be warmer than surrounding rural areas is called the urban heat island effect, and is caused partly by the reduced cooling effect due to evaporation. When Singapore was a rainforest, around 70% of the sun's energy was dissipated through evaporation, but this percentage is now much lower. Cities also get warm because of the huge amount of energy used in the form of electricity and fuel for transportation. Electricity usage of buildings in Singapore ranges from 0.5 to 3.1 Watts per metre square (W/m^2) in small landed properties, to over 1,000 W/m^2 in some of the tallest commercial buildings in Singapore's Central Business District (CBD) (Fig. 16.1). To put these figures into perspective, the average solar radiation received in Singapore is about 180 W/m^2. Thus, many large commercial buildings in CBD use far more energy—and, therefore, produce more heat—than they receive from the sun.

T. Schröpfer and S. Menz, *Dense and Green Building Typologies*, SpringerBriefs in Architectural Design and Technology,
https://doi.org/10.1007/978-981-13-0713-3_16

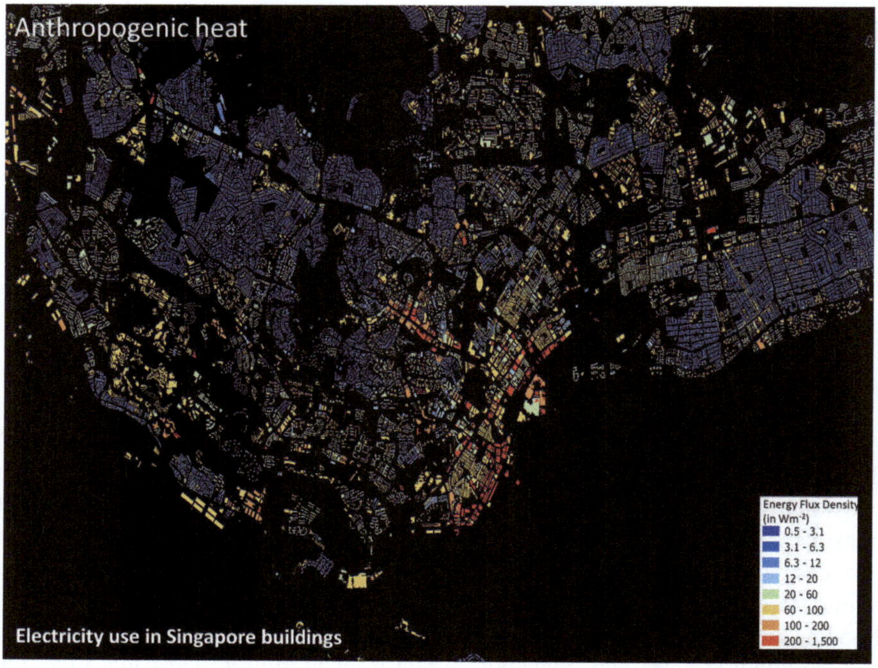

Fig. 16.1 Building electricity use, Singapore; *Credits to* Picture by Dr. Aydt, Heiko; AND Data from Boehme et al. 2015 Sustainable Cities and Society

An analysis of cities around the world by Winston Chow and Matthias Roth of the National University of Singapore showed that the urban heat island effect varies according to geographical location, and increases with the size of the urban population. They recorded an urban heat island effect in Singapore of up to 7° C, with the highest values in the built-up centre of the city. This finding has recently been confirmed using data collected by some 50,000 school children participating in the National Science Experiment, who wore a Wi-Fi-connected multi-sensor device for several days. Unsurprisingly, the CBD—with its very tall buildings—had the highest temperatures, which averaged around 34° C in the early evening, compared with only 26° C in forested areas.

Planners need to take the urban heat island effect seriously, especially in tropical cities, because the problem of high temperatures is likely to increase. A rough calculation suggests that the combined effects of global warming and an increasing urban heat island could raise temperatures in parts of Singapore by as much as 6° C by the end of the century. Given Singapore's hot, humid climate, such an increase would not merely reduce livability, but could even prove fatal for people working outdoors.

Unfortunately, the conventional solutions to the increasing temperature and flooding in cities are often inefficient and in some cases may even aggravate the problem. For example, the conventional solution to avoid flash flooding is to build monsoon storm drains, which, in turn, replace natural vegetation and thereby increase flooding risks. Similarly, the conventional solution to an increasingly warm environment is to use more air conditioning. However, the air-conditioning units in tall buildings produce rising plumes of hot air, which increase atmospheric turbulence and may seed storms. Indeed, the increased turbulence associated with tall buildings may explain why cities often experience more intense rainfall than surrounding rural areas. In Singapore, for example the frequency of storms exceeding 70 mm of rain per hour doubled between 1980 and 2010, a period in which many large buildings were constructed.

These examples illustrate that we need to find more ecological ways of managing the urban environment. Ecosystem services, especially benefits from trees, will certainly play an important part (Fig. 16.2). However, despite growing interest in using greenery to improve the urban environment, we still have rather little precise information about the magnitude of the services provided, especially in the humid tropics. A few studies have been undertaken to investigate the cooling effect of vegetation. These have shown that increased evaporation may reduce urban temperatures by 1 or 2° C, with this effect extending some distance from the vegetated area. Shading by trees is also important, and research shows that the combined effects of shading and increased evaporation by trees may reduce temperatures by between 2 and 6° C. In Singapore, there is ongoing research to studying and quantifying the cooling effects of common trees. One study found that the rain tree produces an average cooling effect equivalent to about 45 W/m^2, which amounts to about one-quarter of net solar radiation in Singapore (180 W/m^2). However, not all tree species are equally effective in cooling. For example, rain trees lose much of their foliage for part of the year, which means that the cooling benefit is also reduced. To get the greatest benefit from ecosystem services, we need more information about the properties of different tree species.

The types of vegetation planted in cities are important (Fig. 16.3). In Singapore, it is usual to grow grass on vacant land lots planned for future land use, presumably because this vegetation is easy to manage. However, is grass actually the best option in terms of ecosystem services? What about other structural types of urban vegetation, such as turf with scattered trees, turf with shrubs, trees with a shrub layer and unmanaged vegetation? Studies are currently in progress to determine how these various vegetation types compare in terms of the services they provide. So far, we only have preliminary results, but we already see big differences for certain services. For example, grass alone is much less effective in preventing runoff than vegetation with shrubs or trees. Therefore, if this particular ecosystem service is important, as it surely is in a city prone to heavy rainfall, short turf is not a good option. Once again, more information concerning the capacity of different types of urban vegetation to deliver ecosystem services is urgently needed as a basis for managing urban species more effectively.

In a densely populated city, much of the land area is occupied by buildings and it becomes important that these also provide ecosystem services. Although interest in green buildings is growing, we know rather little about how much this greenery contributes to environmental quality, especially in the tropics. However, studies in North America have shown that green roofs can reduce ambient air temperatures by

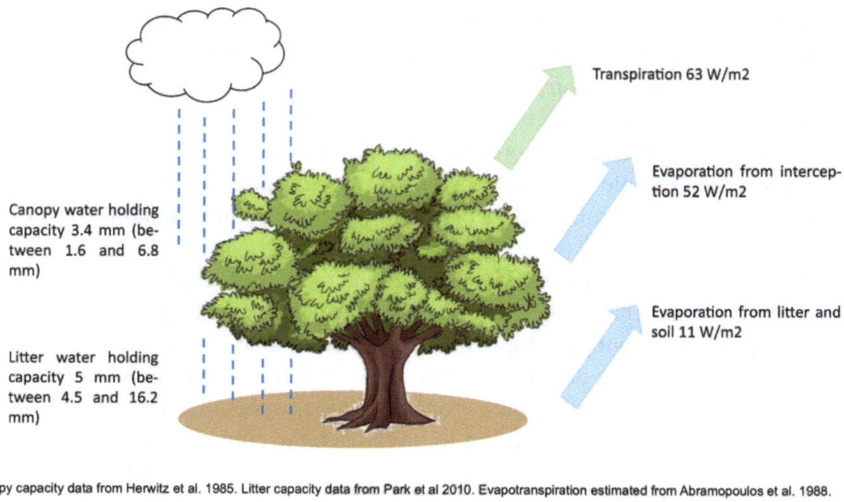

Fig. 16.2 Tree benefits diagram; *Credit to* Peter Edwards

Fig. 16.3 Urban vegetation types, Singapore; *Credit to* Fung Tze Kwan

up to 5 °C, reduce domestic electricity consumption by up to 6%, and storm run-off by up to 90%. Comparable research is urgently needed for the humid tropics.

Let me conclude by making four points on how ecosystem services can be incorporated into planning. The first is that a systems approach is needed in planning ecosystem services. A good example is the Active, Beautiful, Clean Waters (ABC Waters) Programme of Singapore's Public Utilities Board (PUB), which is based upon a detailed understanding of how water flows through an urban catchment. The ABC Programme incorporates a variety of measures for retaining water at the source, thereby ensuring that less water is lost in storm runoff and reducing the need for large storm drains. Taking this idea further, the need to construct storm drains can be thought of as a negative externality of buildings constructed with hard, non-absorbent surfaces. In a sustainable city, such externalities would be internalised, for example by requiring buildings to retain rainwater and therefore not contributing to the risk of floods. The ABC Waters guidelines show many ways that this can be done.

Second, in a land-scarce city, urban landscapes need to be designed to fulfil many functions. In particular, all urban landscapes need to contribute to regulating environmental conditions, rather like the rainforest that existed before the city was built. There are now several good examples of how this can be achieved, such as the Kallang River segment in Bishan Park, Singapore, or Mayesbrook Park to the east of London.

Third, urban landscapes need to be designed so as to strengthen ecological continua. An emerging trend in Singapore and other cities is for the hard boundaries between buildings and the spaces between them to disappear. Buildings are increasingly clothed with greenery, which merge—sometimes imperceptibly—with the vegetation of the surrounding land. This seems to me to be a very positive development, which can help strengthen the contribution of ecosystem services in urban environments.

Finally, new tools and approaches are needed to assist planners and designers in obtaining the greatest value from ecosystem services. Interactive computer tools can allow designers to access ecological information, in much the same way as they can access information about the properties of building materials, and use this information to assess the environmental benefits of particular designs. An exciting new source of information for the designer is point-cloud data obtained from Lidar scanners. This data provides a very accurate three-dimensional representation of a site, including its vegetation, which can be manipulated as part of the design process. Preliminary studies indicate that the resulting point clouds data can be used to quantify the delivery of ecosystem services by particular designs.

I would like to end with my vision of a sustainable tropical city as one that uses ecosystem services to keep cool, to prevent flooding, to provide clean water, and to purify air. Such a city will be less dependent than today's cities upon expensive infrastructure, such as storm drains and large, centralised water-purification facilities. And the buildings will be closely integrated with the surrounding ecosystems, and will themselves be a source of ecosystem services.

Prof. Dr. Peter Edwards has been the Director of the Singapore-ETH Centre since 2013 and is currently the principal investigator of the Ecosystem Services in Urban Landscapes project under the Future Cities Laboratory. In Singapore, he is also an adjunct professor in the Asian School of the Environment at the Nanyang Technological University (NTU). He has been a professor of Plant Ecology at ETH Zürich – the Swiss Federal Institute of Technology in Zürich since 1993, where he has also served as Chairman of the Department of Environmental Systems Science. Prior to this, he was a lecturer/senior lecturer in Ecology at the University of Southampton, England, from 1973-1993. Peter Edwards took the natural science tripos at Cambridge University, specialising in botany, and graduated in 1970. In 1973 he obtained his PhD degree, also from Cambridge, for a thesis entitled Nutrient cycling in a New Guinea montane forest. Peter Edwards has always had a strong interest in the application of science and technology for better policy. He was a founder and first Executive Secretary of the Institute for Ecology and Environmental Management, a professional organisation for environmental practitioners. At ETH Zürich, he was faculty coordinator and member of the Executive Board of the Alliance for Global Sustainability, a research partnership between several leading universities.

Chapter 17
Green Architecture: Landscape Topology and Context

Christophe Girot

'Green architecture' is very new to me. Back in the late 70s and early 80s, architects were working on very elaborate facades. Trees, more often than not, were put on the side in order not to hide the building. Now architects would tend to want to have the trees be the facade itself. That indicates a big level of progress and change in our design culture.

'Topology' is a word that has been stolen from ancient philosophers by mathematicians, but it simply means the knowledge of the ground or study of continual surfaces. We have continual surfaces in our cities that are just made out of asphalt and hard materials. The real question is should they continue being made out of asphalt? Should they continue serving the purpose that they do today or could they be changed? Some cities have been transforming their streets, have changed them into public spaces and into new landscapes. Therefore, topology is a keyword in our understanding of the future city. Basically speaking, what topology is to landscape is what tectonics is to architecture. Architects, urban designers and landscape architects are no longer working in front of a clean slate. The urban context, or more precisely, the urban surface is the result of the functionalist mess that we have created all around. We need to react to the given conditions and cannot reasonably imagine a return to some utopian paradise with almost eight billion inhabitants now on the planet. The idea that we have chopped up our urban space into pieces and created a set of difficult situational conditions has become a constant that has made its way into the deepest recesses of design and fashion.

But architects have always worked on utopias and have always believed in wonderful environmental dreams. The caricature of that dream is, of course, New York City. The urban grid that once landed on the new world as inherited from the Greek, but where the buildings and skyscrapers mushroomed out of control to create something new. One of the main protagonists of such a built utopia was Buckminster Fuller. He operated during the Cold War, at a time when we were not yet talking about climate change but about noxious gases killing humanity. In his utopia, the Empire State Building and the entire Midtown area of Manhattan were to be placed under a giant geodesic dome. During the Expo 67 in Montreal, Fuller created a

T. Schröpfer and S. Menz, *Dense and Green Building Typologies*, SpringerBriefs in Architectural Design and Technology,
https://doi.org/10.1007/978-981-13-0713-3_17

much smaller replica of this geodesic dome, which became a strong symbol of the nascent Whole Earth movement. It embodied the early technical response to many of our environmental concerns manifest in the different ways of living and building in the world depicted in Bernard Rudovsky's book *Architecture Without Architects* published in 1964. So this passage from utopia to the present is something that has already happened and that we should reconsider today, for it is becoming the very basis of a new design culture that is rapidly catching on.

Green architecture started with strong utopian figures. One early *agent provocateur* in that realm was François Roche, who for instance depicted some sceneries of London with Big Ben amidst green towers sinking in a rising sea as a cataclysmic utopia. Of course, each building in this utopia was shown as being self-sufficient, where people were seen growing their tomatoes and raising some chickens somewhere up their sinking towers. At the time it came out, around the turn of the millennium, that idea was very provocative and had a strong impact on the ecological consciousness in architectural circles.

Roche even proposed a design for some form of new Swiss chalet called 'Water Flux' at Evolène in 2009. Placed at an altitude of 2,000 m and dressed with wooden spikes, the building was supposed to catch the falling snow and change appearance over time. It was inspired by the wooden masks that the local population makes each spring for a carnival that is meant to chase people away. The building itself appears like an anthropomorphic hybrid, half object, half being. One can understand it at first as a complete joke, or rather as a mirror of our rapidly changing world.

At the same time, Roche, with his office R&Sie, developed some visionary urbanism with projects like a new administrative town in Korea titled 'Swarm Town' in 2006. It was all generated with organically shaped diagrams using a procedural design method. The 'organic' procedural method seems interesting, until one realises that the programmatic approach remains highly functionalistic, not to say prosaic. Despite the green colour of the 3D model, one can rightly question the true ecological purpose of the project, seeing that it is in fact just a very normal zoning of administrative buildings tucked under some green shell. The content seems to defeat the purpose of this organic urban form.

Roche also worked on a poetic and provocative project called 'Lost in Paris' in 2008. In Paris—a city that has become ecologically correct with some new regulations encouraging green roofs—he proposes a little house addition covered by giant glass bulbs containing noxious bacteria. Some green vines cover the building and make it invisible to the world. The building cannot be seen from the sky, but it is still there.

So what about greening in a historical context like Paris? What about greening an entire city? The utopian 'ecological' projects done for the City of Paris by Belgian architect Vincent Callebaut are the polar opposites of the visionary work done by François Roche. Callebaut brings his architectural touch to Paris in a massive way, his beehive-like architecture is placed like an upside down pudding on top of every building and the resulting dribbling vegetation is indeed very green. But what kind of green are we talking about? Can trees simply be inserted into facades as shown in his illustrations? For any person with any understanding of landscape architecture, it is amazing to see his trees epoxied to the roof, where there is almost no ground provided for them to grow. Apply this concept to the whole of Paris and the city looks like the *Oasia Downtown* building in Singapore to the power of 10. The one most obvious building that should be greened remains un-greened, that is the Eiffel Tower in the distance, just that single fact betrays the irony in the entire project.

Continuing with Callebaut's visionary green Paris, we see an extreme example of ecologically correct indulgence with this eco-bridge to be built in front of one of the chicest quarters of Paris in the 16th Arrondissement. It represents a complete energy-generating green bridge replete with wind-generated propellers for electricity. The project even shows a wet biotope in the foreground with some reeds and a heron flying by. The entire architectural 'salad' shows many ecological conventions uncritically thrown at us, but it somehow does not feel believable in a city like Paris, it seems to belong rather to some amusement park at the periphery of town.

Then there are the serious 'green' cities of the world. Norman Foster pushed the idea of an ecological city near Abu Dhabi named Masdar, which became an experiment in matters of designed ecology and desert greening (Fig. 17.1). If one thinks that Abu Dhabi is known as the second place in the world with the highest CO_2 emissions per capita, trying to make an ecological city there seems not only utopian but very optimistic—especially something so green planted in the middle of an arid desert. It required a positive act of faith demanding massive technical investments in desalination infrastructure and public transportation. Could Masdar actually turn an ecological utopia into reality in one of the most unlikely places on earth? The pictures and design renderings of the promised development remain very appealing; a green town rises from the desert like a mirage, an urban paradise of sorts, and it definitely was marketed as such. But the reality of the built development feels quite different than what was hoped for or expected. Masdar remains rough, not to say arid at best. How far can one go beyond such a 'green' discourse to meet the harder reality of what was actually produced. It will be time, soon enough, to measure the results of the Masdar experiment with the reality of the ecological services that are actually provided but for the time being the utopian dream of a green desert city lives on.

Another visionary project called 'Metrobosco' was conceptualized 10 years ago by the Italian architect Stefano Boeri. It produced an extraordinary reticular landscape for Milan drawing from a regional web of green lines converging on the city centre. The idea was to connect the city with the hinterland of the Alps through a series of green universes ranging from urban to rural. The project includes an extraordinary array of possibilities through green lines and infrastructure that offer a whole list of services.

In logical continuity with the Metrobosco project was the architectural concept of the Vertical Forest which Boeri realised with his emblematic Bosco Verticale project in Milan. The outcome is somewhat disappointing though, with a banal apartment clad with green. The building stands rather lonely in its urban context. It seems obvious in this case that doing a single green building standing out on its own does not really make much sense unless it is connected to a bigger green system and network. This is not a critique on Boeri himself but on the limited impact of his work that provokes an entire set of questions and responses.

Fig. 17.1 Masdar, rendering *Credit to* Foster + Partners

There exists an extraordinary study of the microclimatic potential of vegetation possibilities in cities done by the GeoNet Institute in Hanover, Germany. In their study of the City of Munich, they took the whole city, incorporated the features of its terrain and topography, and did a very thorough analysis of diurnal air movement and temperature. The results were based on thermodynamics and the gravitational pull of the hot air moving out of the city towards the Alps in the late afternoon with the cool air returning down to the city in the middle of the night to refresh the urban environment. The importance not only of vegetation but of its pattern and orientation within the larger territory of a city demonstrated the importance of topological considerations combined with microclimatic optimization.

Cities in temperate climates are now starting to have these so-called tropical nights where the temperature never recedes below 20 °C, which is much too high. Mapping such data against market values for property becomes quite interesting as in the case of the study that was done on Munich. This may completely change the market value of certain places, knowing that you could benefit from cool airstreams coming in other locations. This kind of microclimatic information could also help inform new forms of architecture feeding off the benefits of neighbourhood cooling. In the meantime, other very elegant neighbourhoods in town could end up remaining completely blocked in what we now call urban heat islands.

One sees in the study plan of Munich done by the GeoNet Office that fragmentary gardens and parks, however beautiful they may be, do not really deliver much cooling services to the adjacent neighbourhoods. Topological continuity, flow and connection to the bigger environment are needed in order to properly ventilate a city like Munich. Understanding air temperature and topography, together with the correlations between areas where airstreams happen in relation to the rest of the city illustrate the systemic level of observation that is urgently required in all urban studies today. There is a real urgency for us to learn how to map these cold streams and tap new forms of architecture, landscape and urban design onto them.

One more salient example of infrastructural change for the better is the Cheonggyecheon River project in Seoul, which has become one of the most popular places in town. A huge strip of motorway over a length of 8 km, completed under President Park in the 1960s and financed by the Marshall Plan, was entirely demolished to be turned into an 8 km-long river park. This landscape improvement drastically changed the real estate value in the whole area. The breaking down of the freeway actually alleviated traffic problems in downtown Seoul. There has been some criticism that the new urban river was not ecological enough, but some real herons can be seen here pecking at fish, without even knowing that the water running down the Cheonggyecheon is now being pumped out from the nearby Han River.

Much has been written on the history of the bayous, which were the main natural water bodies running through Houston. Some long-standing plans were created to restore the rivers to what they were into big green fingers that could work as parks. However, the urban reality decided otherwise, and most of the highway infrastructure has either been put on or next to the river. And, as we all saw at the end of the summer in 2017, the whole city was under water because of a mega-hurricane and a failing bayou system that was blocked by the works of man.

The rules should be quite simple with natural tidal waterways like bayous, they need to be continuous and unencumbered, just like roads are kept continuous and free of obstacles. If bayous are neglected or if they are blocked by things, then there can be extremely negative consequences. The famous 100-year flood rule that our engineers and forefathers believed in is now gone and the greening of Houston along the bayous will become a necessary goal in years to come.

The Green Corridor of Singapore obviously cannot flood as a bayou does in Houston but that is not a reason to interrupt its flow with buildings. The corridor could serve as a place for experimentation in diurnal ventilation for some parts of town. It could even become a vital green organ for the City of Singapore providing many added services and values. Therefore, we should not just look at it as a potential site for real estate development but rather look at it organically as a potentially strong physical link between the cold air highlands of Bukit Timah and the warmer harbour area of Tanjong Pagar (Fig. 17.2). We need to study this stretch of land through

Fig. 17.2 Axonometric projection of a point-cloud model along the Rail Corridor in Tanjong Pagar, Singapore; *Credit to*—Philipp Urech

a more systemic analysis. At the Future Cities Lab in Singapore, the Landscape Ecology module is presently studying the possible ecosystem services that could be provided. It is beginning to analyse the Green Corridor through point cloud and hyperspectral technology, looking at the effects of vegetation on air temperature generation and flow, and trying to understand whether these variations will have an impact on thermal comfort if the development of a comprehensive landscape project involving significant urban reforestation will happen.

There is no reason why the Green Corridor could not become a major ecological experiment for Singapore. It is the perfect case study to apply a more sustainable way of thinking to an entire part of town. The conditions in Singapore are obviously fundamentally different than those in Munich. It is a different project altogether, both in terms of urbanisation, climate and scale. Yet people should not move so fast in designing or over crowding the corridor with projects because air flow convection and temperature react to even the slightest gradients in terrain change. One should wait until studies are made to ensure that the expected thermal convection that could effectively happen there unencumbered.

At the other end of the world, in California, it is interesting to note that the two biggest firms in the world, Google and Apple, have been inspired by the early utopias of Buckminster Fuller to create their new headquarters. The Google Headquarters in Mountain View California is designed by Bjarke Ingels Group (BIG) and the building is an accurate reflection of how Google sees the world: the rest of the world may perish but Google will remain protected under its glass dome. So your big data is safe and will be conserved and analysed through and through.

Certainly, the project is shown in an exquisite garden and is all about happiness and beauty. The images of the project show a sort of utopian world becoming reality, where the notions of inside and outside dissolve in a sea of green. This notion of seeing working people being actively involved both within the building and on the outside is quite compelling.

Yet, the most striking part about these green utopian architectures is how self-centred they are. When Steve Jobs asked Norman Foster to design the Apple campus in Cupertino, the idea of a concentric circular building shape came up. Jobs wanted a circle within the circle that enclosed a garden, a sort of 'brain park' right in the middle of the project with a little transparent pavilion at its centre. It was conceived 10 years ago, just before the iPhone came out, and just before the word "selfie" was born. But ultimately this green utopian project of the circle is the first selfie of sorts. It is all about me, myself and I centred on an idealised form of nature.

Again, the notion of opening up to the environment, of exchanging within the community, is a necessity. In that sense, the concentric enclosed shape given to the Apple Headquarters is in a way contradictory to what one should be striving for. The whole Apple campus looks ecological and pristine set in an English Garden style. The streamlined landscape feels almost like a park on the outside, with an exclusive paradise garden on the inside. There is a beautiful think tank in the middle of that park and it is extremely seductive. Yet, there are some serious limitations to this extremely self-indulgent exercise with nature. One should, therefore, question a building that is circular and defensive and that keeps the garden within. Whether both of these built utopias function ecologically or even climatically remains an open question.

To conclude, the ultimate question to 'green' architecture is how it is to be articulated with the rest of the environment and actually how the landscape flows from the gardens within to the outside. As we have seen, landscape topology plays an essential role within the urban context and we should all really start branching out of our buildings to bring a healthier, more comfortable and more energy-efficient world to the forefront of our concerns.

Prof. Dr. Christophe Girot is Professor and Chair of Landscape Architecture of the ETH (Swiss Federal Institute of Technology) in Zürich since 2001. He was born 1957 in Paris France. He holds a Bachelor of Science in Environmental Planning from UC Davis, California 1981 as well as a Masters of Architecture 1986 and Masters of Landscape Architecture 1988 from UC Berkeley, California. He was Chairman of Landscape Architecture and Design at the ENSP Versailles 1990-2000. He will be the next Dean of the ETH Department of Architecture for the period 2017-2022. He directs the ETH Institute of Landscape Architecture since 2005 and the ETH LVML (Landscape Modeling and Visualizing Laboratory) since 2009. His emphasis in research and teaching is on large-scale landscape design and modelling with particular attention to the topology of natural systems in cities. This has enabled significant ecological design improvements through advanced point cloud modelling techniques. He holds a practice in Zürich with projects that test the current limits of topological design and modelling in extreme alpine situations. Christophe Girot has published extensively and just completed a major book entitled "The Course of Landscape Architecture" published by Thames and Hudson in 2016.

MIX
Papier aus verantwortungsvollen Quellen
Paper from responsible sources
FSC® C105338

If you have any concerns about our products,
you can contact us on
ProductSafety@springernature.com

In case Publisher is established outside the EU,
the EU authorized representative is:
Springer Nature Customer Service Center GmbH
Europaplatz 3, 69115 Heidelberg, Germany

Printed by Libri Plureos GmbH
in Hamburg, Germany